IMPACT PHENOMENA
IN TEXTILES

IMPACT PHENOMENA IN TEXTILES

PUBLISHED 1963 BY THE M.I.T. PRESS, CAMBRIDGE, MASSACHUSETTS

W. JAMES LYONS

FOREWORD

There has long been a need in science and engineering for systematic publication of research studies larger in scope than a journal article but less ambitious than a finished book. Much valuable work of this kind is now published only in a semiprivate way, perhaps as a laboratory report, and so may not find its proper place in the literature of the field. The present contribution is the nineteenth of the M. I. T. Press Research Monographs, which we hope will make selected timely and important research studies readily accessible to libraries and to the independent worker.

J. A. Stratton

PREFACE

The aim of the author in the preparation of this monograph has been to provide, under one cover, an account of the developments that have taken place in the study of impact phenomena in textiles over the past half-century. Notice has been taken of all material, pertinent to the subject that has come to light and been made available. At the same time, the possible utility of the various items of information has been the criterion for determining the depth with which they are treated. Use of the book not only by scientists with advanced knowledge of mathematics and the theories of elasticity, rheology, and molecular mechanisms, but also by technical people of limited training in these particular fields, has been anticipated (successfully, it is hoped).

The methods and apparatus of impact research on textiles (Chapter 3) have been covered quite thoroughly, in that mention at least has been made of each one to come to the author's attention. It is believed that sufficient descriptions are given, and the pertinent principles stated, so as to be fully understandable to an experienced research scientist or instrument engineer, in all important respects. The results of studies of impact phenomena (Chapter 6) that have been included have been selected to cover a broad spectrum of fiber types, with emphasis on those of modern or potential interest in the mechanical field.

Repetitious results, merely confirming earlier findings, have been, for the most part, omitted. The volume of material in the literature has prohibited incorporating within the scope of this treatise all the reported experimental details. For such items, the reader is referred to the publications cited.

In the initial phases of the preparation of the References, the author had the assistance of William P. Virgin. Mr. Virgin also prepared rough drafts of parts of Chapters 2 and 3. The author is happy to acknowledge the cooperation of Roy C. Laible, Project Officer representing the Quartermaster Research and Engineering Command during most of the contract period. Mr. Laible was very helpful in providing Government reports and other material used in the present work. The author has also had the benefits of the comments and encouragement of Dr. John H. Dillon, who read the original manuscript as the successive chapters were completed.

Textile Research Institute   W. James Lyons
Princeton, New Jersey
January, 1962

PREFACE TO THE PRESENT EDITION

This monograph, prepared at Textile Research Institute under the sponsorship of the Quartermaster Research and Engineering Command, U.S. Army, first appeared in March 1962 as a report to that agency. Since the completion of that report, further developments in the study of the impact properties of textiles have, of course, occurred or have been reported. Cognizance of these developments has been taken in the present, revised edition, with appropriate expansions in the text and additions to the References.

The author wishes to thank the following organizations, journals, and individuals for granting permission for the reproduction of illustrations previously published elsewhere, or for furnishing original photographs, and in some cases, for both: Transactions, American Society of Mechanical Engineers; American Society for Testing and Materials; Journal of Applied Physics; Professor Stanley Backer, Massachusetts Institute of Technology; Celanese Corporation of America; Cook Electric Company; Fabric Research Laboratories, Inc.; Dr. Douglas W. Hill, Shirley Institute, Great Britain; Institution of Mechanical Engineers, Great Britain; Professor W. R. Lang, Gordon Institute of Technology, Australia; Textile Institute, Great Britain; Textile Research Journal; and John Wiley & Sons, Inc.

Use has been made of illustrative material, cleared for public release, which originated in, or was produced under the sponsorship of one or another of the following U.S. Government agencies: Aberdeen Proving Ground, U.S. Army; Aeronautical Systems Division, U.S. Air Force; Chemical Research and Development Laboratories, U.S. Army; National Bureau of Standards; and Quartermaster Research and Engineering Command, U.S. Army.

It is a pleasure to record also, and with thanks, that the cooperation and encouragement of Dr. S. J. Kennedy, Mr. Louis I. Weiner, and Dr. W. E. C. Yelland, of the Quartermaster Command, have facilitated publication of this edition.

Princeton, New Jersey   W. J. L.
February, 1963

CONTENTS

Chapter 1

INTRODUCTION

1.1 Textile Applications Involving Impact

In many mechanical applications, textile materials are subjected to rates of loading or straining that are many times greater than those encountered in ordinary apparel or household uses. It has been estimated, for instance, that in a regular passenger-car tire, traveling at 50 mi/hr, the cords are repeatedly strained at rates that are equivalent to sinusoidal frequencies in the order of hundreds of cycles/sec.[49] In terms of average extension, the rate approaches, or may even exceed 1000 %/sec, depending on the constitution of the fabric and the conditions of use. Calculations indicate that in the brief interval before the opening of a military parachute the suspension lines are subjected to an average rate of straining of about 250 %/sec, with an attendant force of about 1 ton. Modern high-speed stitching, in the commercial manufacture of textile articles of all kinds, subjects the thread to impact velocities which may be equivalent to even greater rates of extension. Stone, Schiefer, and Fox[104] mention velocities ranging from 1 to 10 m/sec being imposed on thread as often as 5000 times/min. It can be readily deduced that the corresponding accelerations are in the order of 10^6 cm/sec^2. Calculations of the present author, based on an analysis of the high-speed stitching operation by Morgan,[69] indicate that the accelerations of the thread can reach, at least 3.6 $\times$ 10^6 cm/sec^2.

The behavior of textile materials under conditions of impact or shock loading is important not only in the applications mentioned above (pneumatic tires on highway vehicles and off-the-road equipment, and parachute suspension lines), but in many other uses. In the military supply system are such items as armor clothing and climbing ropes, which depend for their adequate performance on the ability of the component yarns to withstand the effects of shock loading.

The advent of the airplane, with the attendant high velocities (especially in military operations) brought new and unprecedented problems in areas where textile structures were put to use. Aircraft tires during landing are subjected to severe impact loads at the instant of touchdown. The military use of aircraft has extended, in the last quarter-century, to the aerial delivery, by

1

retarded fall, of combat personnel (paratroopers), arms, and supplies. This, in turn, has led to the development of a broad technology in the design, construction, and use of parachutes, which, with auxiliary equipment, are almost wholly textile structures. Consideration must be given to the ability of such components as canopies, static lines, and harness, to absorb the energy of impact loading, especially during deployment and opening of a parachute. Another aeronautical application of the parachute, where again the impact properties of textile components are important, is in the deceleration of high-speed aircraft during landing.

Among the industrial applications of textiles where impact behavior is the crucial property are the safety belts and lines used by construction workers, and the heavy ropes used in the marine, and other, industries.

These expanded, modern uses of textiles have brought with them the realization that the performance of the materials in these structures depends upon a set of properties quite different from those that determine the performance of textiles in their more familiar uses. It became apparent that these "impact" properties, as they came to be called, could not be reliably deduced or inferred from the behavior of the materials under relatively slow application of load. The evaluation, comparison, and analysis of these properties, accordingly, call for test methods and theoretical concepts different from those conventionally employed. As Leaderman aptly says, "There is no justification for evaluating the fatigue behavior of a tire-cord or the resistance to shock load of a parachute fabric by means of the usual standard tensile test. . . ."[46] Thus, a number of experimental techniques have been developed for studying the impact properties of textile structures in various ranges of strain rate.

1.2 Early Studies

While technological developments in the last twenty years have stimulated interest in the investigation of the properties of textiles at high rates of strain, the desirability of testing such materials by an impact method was recognized a half century ago. In 1910 Lester[47] described a "ballistic" method for measuring the work of rupture, employing a falling pendulum to which one end of the specimen was attached. Results on a cotton sewing thread were given. Lester was evidently interested in the technique principally as a better way of obtaining the work of rupture, than the tedious computation from stress-strain diagrams.*

---

*Of incidental interest, and surprise, considering that it was made as late as 1910, is Lester's remark that ". . . it has never been customary, as far as I am aware, to make stress and strain diagrams from specimens of textiles. Nor is the writer prepared even to suggest that the preparation of such diagrams should be encouraged. . . ."

However, he concludes, among other things, that: "Since the method gives proper importance to the elasticity of the sample, it indicates better than existing methods the true 'strength' of the yarn, the 'toughness' and the actual working quality.   It gives the amount of resistance of yarn to the shocks of weaving, and it gives the resistance of cloth to those strains which most commonly cause it to fail in use."

Five years later Balls, in his early book,[10] described a pendulum apparatus, inspired by the Lester article, for impact tests on bundles of cotton fibers. About this time, interest was developing in the adaptation of impact pendulums to the testing of fabrics. Pickard and Wallace in 1919[81] described an instrument for measuring the impact resistance of fabrics, which eliminated a defect in design in an earlier pendulum machine having the same purpose. A description of the latter instrument was not published until about ten years later, when it was used (without modification) by Morton and Turner[71] in a comparison of various strength tests on fabrics. Both teams concluded that, because of the variability in results between specimens of the same sample, and for other reasons, the impact test was unsuitable for evaluating fabrics. The interest in impact tests during this early stage continued to be in the quality, convenience, and speed of these tests, as replacements for dead-load methods.   The importance of the fact that the two types of test measure material properties under significantly different conditions was only dimly recognized.   However, Morton and Turner did record their opinion "that fabrics in general use are not ordinarily subject to impulsive forces of the type operative in the impact tester, so that _a priori_ there is no real application for a test of this kind." Summing up the situation during this period, Balls, in his later book,[11] says: "The impact test measures the energy absorbed in the act of breaking the yarn, and it has great advantages of convenience.   Yet it has been repeatedly advocated and discarded by students of textiles."   In this book he describes another pendulum device for impact tests on yarns. Making the point that the impact test is distinctive, Balls says ". . . the Time Factor intervenes, so that dead-load measurements have no exact relation to the reality of instant breakage."

There was interest in this factor, the effect of the rate of loading on properties of fabrics used in early aeronautical craft, going back as far as 1910.[12] The times to rupture involved in these studies, however, were in a range corresponding to slow quasi-static testing. Recalling Lester's pioneering work, Denham and Brash,[25] in 1924, described in some detail an impact pendulum for "ballistic" measurements,* and presented an imposing

---

* In earlier years, use of a falling pendulum was quite commonly called a "ballistic" method, because, no doubt, it involved load application that was "instantaneous" in comparison with con-

array of data on silk yarns, intended to explore the reproducibility of the technique, and the effects of varying certain experimental parameters. In the same year Foster[32] published some
energies of rupture of bundles of cotton fibers, measured on an
undescribed pendulum impact tester.

In 1926 F. T. Peirce, singly and in collaboration with others,
published a series of papers on the tensile testing of fibers and
yarns, in which the role of the "time factor," or rate of loading,
was comprehensively explored. In the first paper, Mann and
Peirce[54] described experiments on cotton fibers in which the rate
of loading (g/sec) was varied over a 1000-fold range, but all of
the tests were conducted on what were essentially quasi-static
instruments, the highest rate of loading (3.57 g/sec) being equivalent to an average velocity of about 0.047 cm/sec.   A falling
pendulum for the impact testing of yarns, built and operated along
the lines of the Lester instrument, was described in a subsequent
article by Midgley and Peirce,[66] who also gave the results of a
number of experiments with the instrument.   In a companion
article, the same authors[67] reported on the effect of the rate of
loading on results obtained by static and quasi-static methods
as well as by the impact method, using their pendulum apparatus.   In the last of the papers involving the rate of loading,
Peirce[75,76] presented the theoretical dynamics of three types of
tensile testing instruments, including the impact pendulum, and
discussed the effects of time or rate of loading in experiments
on various testers.

With the work of Peirce and his colleagues, the theory and
practice of impact testing by means of the pendulum may be
considered to have been brought to maturity.   The explanation
given for one of the findings in this research, namely, that the
work of rupture is not exactly proportional to the length of the
specimen, led to a development which has important bearing on
the statistical theory of the testing, not only of textiles, but of
materials generally.   The same result was obtained by Denham
and Brash, who made no comment, and before them, by Lester,
who found the "discrepancy" inexplicable.   Midgley and Peirce
ascribed this result to irregularity in strength along a yarn, the
time required to stretch a yarn to rupture, and an energy factor.
Peirce, in a separate article, applied the idea of irregularity to
conventional, breaking-test data, to develop the statistical
"weakest-link" theory, for which he has become well-nigh famous.

A search of the literature discloses that little new work was
done on the impact testing of textiles during the next decade, in

---

ventional practice. The term, however, has become misleading,
since there are now methods, employing free-flying missiles,
with which "ballistic" is more appropriately associated.

the 1930's. Evidently, impact testing according to a set procedure had become common practice in some laboratories, for
Peirce mentions the ballistic tester as being, in 1927, "now a
standard commercial instrument."[76] Such a machine was the
Goodbrand Ballistic Tester,[33] of the pendulum type, used by Turner
and Venkataraman early in this period, in a comprehensive study
of yarns from standard Indian cottons.[109] While developments continued in the impact testing of metals and, to a lesser extent, of
plastics,[85] at really high speeds, what interest there was in the
rate of loading of textiles was limited to speeds in the quasi-
static region.[13,26] Leaderman mentions work published in Russia
during this period, on the impact testing of bast fibers.[41]

1.3 Gravity Instruments Since 1940

With the onset of World War II, there was renewed awareness
that many textile items in military use were subjected to shock
loading, and that in some cases, human survival depended on the
proper functioning of the items. There was a corresponding renewed interest in the impact testing of textiles.   A relatively
massive machine, described by Leaderman[46] and Schwarz,[89] was
installed at the Massachusetts Institute of Technology, for the tensile impact testing of heavy linear specimens.   This apparatus
employs the vertical, falling-weight principle, and was equipped
with an electrical strain gage,[36] so that the instantaneous tension
in the specimen during rupture could be obtained. Smith, in his
1944 Marburg Lecture, gave a force-extension curve for nylon,
obtained at 21 ft/sec with this machine [Reference 90 (pp. 36-37)].
Dickson and Davieau[27] have described three impact testers of the
pendulum type constructed in 1943 and in the following years.
These instruments, of various capacities, ranging up to 67.5
ft-lb, were used for impact tearing and tensile tests on a wide
variety of fabrics, as well as for impact tensile tests on yarns.
In 1945 to 1946 Newman and colleagues[73,103] described equipment, of the falling-weight type, to measure the impact rupture
properties of nylon and sisal ropes and of parachute webbing. A
few years later, Wegener, in a series of papers,[116] reported studies of stress-strain-time relationships in rayons, in which also
the falling-weight principle was employed. These were followed
by a paper by him and Geuthe[117] in which the free-fall method was
compared with the constant-rate-of-loading method for determining the stress-extension curves of silk and synthetic fibers.
In 1953 Schiefer, Appel, Krasny, and Richey[86] described a falling-
weight apparatus and presented the impact properties of a wide
variety of yarns and monofils of man-made materials. In all of
these free-fall devices, except that of Wegener, the test specimen, mounted vertically, supported a metal plate or rod.   Impact loading was achieved by allowing a weight to drop through a
known distance on to the plate or rod.

There evidently had been further developments at the Shirley Institute on the falling pendulum of Midgley and Peirce, mentioned earlier in this chapter. In the British Standards Handbook of 1949,[15] specifications are given for tear testing with a modified version of this instrument, in which the single pendulum is replaced by two pendulums. The second, light-weight pendulum is used to hold one end of the specimen, preparatory to impact by engagement with the other heavier pendulum at the bottom of its swing. Thus, as in the falling-weight devices, the specimen remains stationary until the instant of impact. During the early postwar period, other new designs, aimed at improving the falling-pendulum techniques, were reported. Thus, Meredith,[62] in a study of the effect of the rate of extension on the load-extension properties of cotton yarns, used a pendulum apparatus on which the stationary clamp was supported through a cantilever strain-gage. He was thus able to obtain, by means of a drum camera, a time record of the load on the specimen during the impact process. About the same time, Lang,[45] described equipment and methods for adapting the falling-pendulum technique to the impact testing of very short textile specimens (0.1 in. minimum).

The accumulation of data on a wide variety of structures and materials, which these experiments of the 1940's and early 1950's provided, gradually established the fact that there is no universal relationship between the quasi-static and the impact, mechanical properties of textiles. Attempts to devise formulas by which some impact property could be estimated from quasi-static data, disclosed exceptions beyond limited ranges of structures or materials. In some cases, exceptions fitted a pattern, but from this little better than <u>a posteriori</u> rules covering the exceptions could be devised.

The lack of quantitative correlation between low-speed and impact properties was further brought out in the work of Lyons and Prettyman.[51] They used, on a variety of tirecords, a pendulum tester substantially of the original type of Midgley and Peirce, but with the specimen differently mounted (not affixed to the pendulum). The specimen was held so as to form a "V" in a horizontal plane, so that it was struck by the pendulum bob at the apex of the "V." An improvement in the technique was the use of a high-speed camera to follow the impact rupture process photographically. By this means measurements were not limited to rupture energy, but were extended to loading force and corresponding elongation.

More recently Parker and Kemic[31,83] and Ballou and Roetling[9] have described equipment for the impact testing of textiles, also employing the falling pendulum as the loading device. In both apparatus the specimen is mounted in conjunction with load and elongation transducers, so that by means of electronic circuitry,

the stress-strain curve is presented as a trace on a cathode-ray
oscilloscope. While the speed of impact in the Ballou and Roetling apparatus is rather modest (2.12 m/sec), an average strain
rate in the order of 2000 %/sec is obtained by the use of a short
(5-in.) specimen.

During recent years (1955 to 1960) research on the impact behavior of textile structures has been actively revived at M.I.T.,[55,56,57]
using the falling-weight instrument mentioned above, for the experimental phases of the studies. This machine has been modified so that heavier textile assemblies, such as webbings, can be
tested. In addition to improvements in the design of the machine
for safety and facility of operation, there were installed an extensometer employing magnetic tape, and a piezoelectric gage
for the measurement of instantaneous load on the specimen. With
this equipment impact measurements have been made on yarn,
thread, cord and webbing (principally of nylon).

A falling-weight device, mounted in a tall, outdoor, structural-
steel frame has recently been used in work done for the du Pont
Company[29] to compare the impact energy-absorbing properties
of nylon, Dacron, manila, and steel-wire ropes.   Provisions
were made for measuring the tensile force on the specimen during impact, by means of a load cell, from which the specimen
was suspended. The extension of the specimen to break was accurately traced on an oscillograph by means of a driving connection between the falling weight and a potentiometer.

1.4 Methods Employing Nongravitational Forces

The velocities of impact obtainable with pendulums or falling
weights are limited by the height available in laboratories from
which the impacting mass may be dropped. To obtain very high
impact velocities, comparable to those found in some textile operations or applications, other means have been employed to provide the impacting force.* In the early 1950's Meredith[63] designed
an instrument with which extension rates up to about 1100 %/sec
were obtained, in a study of the strain-rate effect on the rupture
properties of various yarns. The extension mechanism consists
essentially of a rotating disc carrying a projection.   When a
steady, high speed has been attained, this projection is made to
engage a clamp holding one end of the specimen. By means of an
optical lever, the extension as a function of time is recorded on
photographic paper in a drum camera.

At about the same time, Schiefer and colleagues[60,104] started a
program of studies on the very rapid loading of yarns, also in-

---

*A mechanical lever, having unequal arms, would evidently
serve as a means of indefinitely multiplying the impact velocity
of a gravity device. This scheme, however, appears never to
have been used.

volving, in its initial phase, the use of a rotating-disc machine. While the specimen is impacted longitudinally, as in the Meredith apparatus, their equipment is much more massive. A high-speed camera, with a system of mirrors, is used to obtain a photographic record of the motion of the ends of the specimen. From this record, stress-strain relationships are computed. Results have been obtained on a number of synthetic yarns at impact velocities ranging from 20 to 200 ft/sec with corresponding rates of extension ranging up to 4×10^5 %/min.[87]

A rotating-disc machine was developed by Parker and Kemic[31,74] for the impact testing of tirecord transversely, rather than longitudinally, as has been the almost universal practice. With a rather ingenious technique, they measured rupture energies at impact velocities up to 80 ft/sec.

For independent studies of recovery phenomena in single fibers, Wood and Chamberlain[119] and Cross and Garrett[23] have devised instruments for rapidly loading comparatively short specimens to predetermined levels, below the rupture point. These machines, embodying much electronic instrumentation, provide rates of extension in the range 10 to 1000 %/sec. In the researches to which these devices were devoted, the interest was primarily in the recovery of fibers after rapid extension; no intentional provision was made for a detailed study of the impact process itself.

To attain higher velocities of impact, in the order of those of missiles in military combat, recourse has been taken to ballistic methods. A suitable mass is projected into free flight, and thus strikes the test specimen. In all the adaptations of these methods to textiles, except one, the specimen yarns or fabrics have been impacted transversely. A description of the operating procedure involving such a ballistic technique for the acceptance testing of armor vests, and their protective fabrics, was issued by the U.S. Army, Aberdeen Proving Ground,[111] in 1953. In this (frequently called) "V_{50}" test, 0.22-caliber steel missiles of special shape are fired, by means of an explosive powder, from a rigidly mounted rifle, at a panel of the fabric system under test. Means are provided for measuring the velocities of the impinging missiles. By altering the powder charge in the rifle the operator can obtain various velocities. The criterion of ballistic resistance is taken as the mean of the velocities, within a specified narrow range, at which 50% of the projectiles are, in effect, stopped by the test panel.

Impact by free-flying missiles is similarly used by the U.S. Navy for the evaluation of the resistance of textile and other armor materials, in a technique reported to have been developed in the early 1940's.* In this "forward fragment spray" test, a

____

* Private communication, C. B. Green, U.S. Naval Weapons Laboratory, 17 Nov. 1960.

panel of the material to be evaluated is placed 3 ft behind a $\frac{1}{8}$-in. mild steel plate, in the line of fire. An explosive-loaded projectile is fired at the steel plate, which, on being punctured, detonates the projectile charge; the resulting projectile fragments retain the forward motion and, as a spray, impact the test panel. The measure of resistance to impact by fragments is the average number of complete penetrations for a minimum of five specimen panels.

The ballistic acceptance method of the U.S. Army, and modifications of it, employing a missile in free flight, have been adapted to research purposes, and have become the basic elements of elaborate electronic systems in a number of laboratories. Petterson, Stewart, Odell, and Maheux[78,80] have reported a study of strain in single yarns under high-speed impact, in which a gun powered by compressed helium was used to attain a range of high impact velocities up to 1900 ft/sec. By means of high-speed, flash photography and other novel, optical techniques,[39,40] records of the propagation of strain in yarns impacted transversely were obtained. Similar equipment was developed about the same time by Coskren and Morgan at Fabric Research Laboratories.[5,31] A contribution of the research in which this equipment was used was an analysis of the part played by aerodynamic drag in the deformation of an impacted yarn — an effect not previously considered.

Supplementing their longitudinal-impact studies with the rotating-disc machine, Schiefer and colleagues[100] have used a spring-driven device for projecting a missile into free flight, in transverse-impact tests on nylon, Fortisan, and Fiberglas yarns. With this apparatus, missile impact velocities up to 225 ft/sec have been obtained. By means of high-speed photography the displacement of the yarn, and its sequential configurations immediately following impact, were recorded. Lewis[48] and Holden[37] have recently described apparatus in which a missile fired from an air rifle impacts a pivoted arm, and thereby rapidly strains longitudinally a yarn specimen attached to the arm. Through electronic transducers and circuitry, stress and strain are registered simultaneously on a cathode-ray oscilloscope, and recorded photographically. The rupture properties of yarns of a number of man-made fibers were obtained at rates of extension ranging up to 66,000 %/sec. More recently, Hall[35] has used this equipment and that of Meredith, to study strain-rate effects in oriented isotactic polypropylene monofils, through a six-decade range of strain rates, up to 49,000 %/sec.

Rocket-powered sleds, of the type used in aeronautical research to simulate on the ground the high-velocity conditions met with by equipment in flight, have been adapted to the impact testing of heavy textile structures, both in Britain[14] and in the United

States.[118] In the latter case, by means of telemetry and other electronic circuitry, data on the breaking loads and energy absorption of nylon webbings were obtained at velocities up to 750 ft/sec.

Recent years have witnessed the analysis and experimental study, by Morgan and others,[70] of the dynamic behavior of yarns and threads in high-speed textile processes, such as weaving and stitching. Rounding out a half-century of study of the impact properties of textiles, development and exploration continue on new devices for the very rapid straining of specimens. Most of these later developments thus far described in the literature, depend on pneumatic pressure, directly or indirectly, to load the specimens. Such are the machines described by Supnik and Silberberg,[105] Krizik, Mellen, and Backer,[42] and by Laible and Morgan,[43] in which tension is rapidly applied to the specimen by the action of compressed gas on a piston to which the specimen is attached. While capable of applications to a variety of materials, these devices have been used on yarns to achieve rates of elongation up to 10,000 in./min. In another investigation, Chu, Coskren, and Morgan[18] have participated in the design of a system of apparatus for the impact testing of heavy textile structures such as webbing, having quasi-static breaking loads in the order of tons. The impacting force is provided by a free-flying missile propelled from a gas gun employing either nitrogen or helium. Velocities up to 800 ft/sec are attainable with this equipment. In all of these testing assemblies, as in others developed in recent years (mentioned in preceding paragraphs), ample use has been made of electronic instruments to measure the pertinent quantities, and of photography to record the results.

A pneumatic instrument for single fibers is in use at the laboratories of the Celanese Corporation of America.[70]   At the Quartermaster Research and Engineering Center, U.S. Army, a "sling-shot" machine powered by heavy rubber bands, for the longitudinal impacting of yarns, has been developed.[82]

Work continues at a number of other laboratories — the most prolific group being that at the National Bureau of Standards, under J. C. Smith. This group has recently published experimental results, obtained on yarns of an extensive range of fiber types, at low speeds of loading, as well as under impact conditions.[101] Still more recently has come a study of the effect of yarn structure on impact behavior.[93]   To the techniques being used has been added a ballistic method[95] along the lines of those pioneered by the Chemical Warfare Laboratories. Coskren, Chu, and Morgan during 1962 reported a number of results on the behavior of parachute components loaded at impact velocities up to 700 ft/sec.[19,20,21]

The advancement of the theory of impact phenomena is not being neglected either. Morgan has contributed to a general treatise on high-speed testing, a discussion of impact behavior as a stress-relaxation phenomenon, for the interpretation of results obtained on textile samples. Much new theory and mathematical analysis appear in other recent publications of Smith and associates, including the calculation of strain-wave velocities,[91] shock-wave effects,[92] and the relation of "critical" impact velocity to the tenacity-extension curve of a yarn.[114] An effect long known to exist, that of air drag on the motion of a filament impacted transversely, has come under analysis in a very recent paper by McCrackin.[59]

Chapter 2

MECHANICAL PROPERTIES OF MATERIALS

2.1 Introduction

The following sections are presented to clarify for those readers who are not fully conversant with the technical aspects of the subject, the basic concepts and terms used in delineating the deformational behavior and properties of matter. The contents are intended to provide some background for the more detailed analysis of phenomena at high rates of straining given in later chapters.

2.2 Stress, Strain, and Hooke's Law

2.2.1 Stress and Tenacity.

When external forces act upon a body so as to change its size or shape, internal forces develop in the body to resist the change. Stress is such an internal reaction expressed as force per unit area.* Stress is to be regarded as acting at every point in the area with which it is associated, though it may not be uniform over the whole area. The stress at a point, then, is defined as the force acting on an infinitesimally small, elementary area including the point, divided by the magnitude of the area. An average, or mean stress, which is frequently used, on the assumption that the stress is uniform over the whole area of a specimen, is given by the relationship:

$$\sigma = \frac{F}{a} \tag{2.1}$$

where σ denotes the stress, F is the applied force, and a is the area over which the force acts. Thus, consider two bars each of which is subjected to a stretching force or load of 1 g. Let one bar have a cross-sectional area of 1 cm^2 and the other an area of 2 cm^2. Then the (mean) stress in the first bar is $\sigma = 1$ g/cm^2 and in the second bar, $\sigma = 0.5$ g/cm^2.

Tensile stress, called tension; compressive stress, called pressure; and shearing stress exist in a body according to the way in which the external forces tend to deform the body. If the external forces tend to increase the length of the body,

---

* The word stress is sometimes used to denote the condition in a body, but it should not be confused quantitatively (or dimensionally) with force alone.

12

such as when they act along the same line away from boundaries of the body, the induced stress is tension.   Thus, in the
bar shown schematically in Figure 2.1, there is acting across
any section AB a tension such that the pulling forces P acting on
the ends of the two parts are equal in magnitude but opposite in
direction.   If the forces acting on the ends of the bar — or of
either part — are 100 g, then the tension at AB is 100 g divided
by the cross-sectional area of AB.

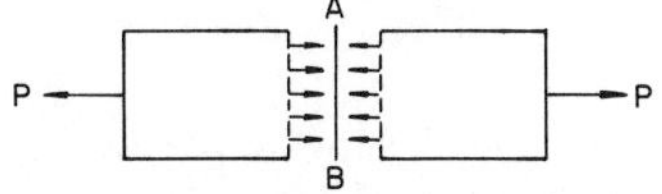

Fig. 2.1. Forces acting across a
section of a bar under tension.

Pressure is developed in a body when the external forces,
acting along the same line toward the boundaries of the body,
tend to decrease the length in the direction in which they act.
In the case of a bar, compressive forces acting on the ends
would tend to shorten it.   Shearing stress is developed in a
body when a pair of equal and opposite external forces, not in
the same straight line, act on the body in such a way that adjacent parts tend to slide over one another.   Compressive and
shearing stresses play minor roles in the mechanics of textiles.

In textile technology the internal forces which are induced in
a yarn or filament, tending to oppose an increase in length when
external stretching forces are applied at the ends, are referred
to the linear density (fineness) of the specimen rather than the
cross-sectional area.   Linear density is the mass (less rigor-
ously, the weight) of material in some standard length.*   The
condition of tensile stress in a textile specimen, expressed with
reference to linear density, is called tenacity.   In textile yarn-
numbering systems, tenacity is given in g (of force)/den or g/tex.

2.2.2 Strain.   The relative change in position in parts of a body
as a result of the application of forces, is called strain.   Thus,
the ratio of the change in length of a bar (either an increase or a
decrease) to the original unloaded length is, in most applications,
taken to be the strain in the bar.   Symbolically, tensile strain or
extension is given by the relationship:

---

*In the C.G.S. (centimeter-gram-second) system of units, this
standard is taken as unit length, and linear density is expressed
as g/cm; in the denier system of yarn numbering, the gram is
also used as the unit of mass, but the standard length is taken as
9000 m, and the linear density is expressed in deniers;   in the
new tex system, the gram is also used, and the standard length
is taken as 1 kilometer (1000 m), the linear density being expressed in tex.   The following relationships hold between these
units: $1\,\text{g/cm} = 9 \times 10^5$ den $= 10^5$ tex.

$$\epsilon = \frac{\ell - \ell_0}{\ell_0} \tag{2.2}$$

where ℓ is the strained (total) length, and ℓ_0 the original length of the specimen. From the rigorous viewpoint, ϵ, like σ, is a mean, but it is frequently used to characterize the condition at every point in a body, on the assumption that the strain is homogeneous.* As an example of the application of Equation 2.2, suppose that a yarn 10 in. long is stretched until it is 11.5 in. long. Then the extension is

$$\epsilon = \frac{11.5 - 10}{10} = 0.15$$

which would, most frequently, be expressed as 15%.

When ℓ in Equation 2.2 is smaller than ℓ_0, ϵ is negative, and the strain is <u>compression</u>. Still another type of strain is the <u>shear</u>, associated with shearing stress, mentioned earlier in this section.

   2.2.3 Stress-Strain Relationships.   That a solid body may be distorted by the application of external forces has undoubtedly been known in qualitative fashion since antiquity. It is widely recognized, furthermore, that if the applied forces are not too great the body will return to its original configuration when these forces are removed. The property by virtue of which a body recovers its original size and shape is called <u>elasticity</u>. The upper limit of the range of stress in which the body exhibits perfect elasticity is termed the <u>elastic limit</u>.

   The general relationship between the stresses and strains that develop in a body when it is loaded is shown graphically in Figure 2.2. With most elastic materials, when successively higher and higher tensile forces are applied to a body, the extensions which first occur are proportional to the forces (or stresses). This observation was first made by Robert Hooke. His statement of this relationship, in 1678, is known as <u>Hooke's law</u>.   In Figure 2.2 the proportionality between stress and extension is represented by the straight segment OA' of the curve. The stress OA above which the curve departs from its straight-line character is called the <u>proportional limit</u>.

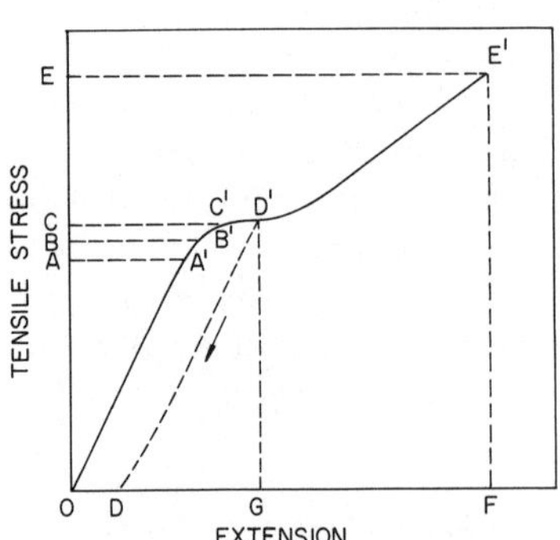

Fig. 2.2. Generalized tensile stress-extension curve (polymeric materials).

---

   *In the consideration of strain-propagation effects in Chapter 4, there arises the situation where the strain is not the same at every point in a yarn, at every instant.

   Laws similar to Hooke's have been found to hold for the other
types of strain, as well as for extension. Hence, the generalized
Hooke's law states that the stress associated with any strain in
a body is proportional to the strain. Expressed mathematically,
this statement is: Stress equals a proportionality constant times
the strain. For the tensile case, with which we shall concern
ourselves solely:

$$\sigma = E \times \epsilon \tag{2.3}$$

where E represents the proportionality constant, called <u>Young's
modulus</u> or the <u>stretch modulus</u>.   Since ϵ is a dimensionless
quantity, being simply a ratio of two lengths, E is expressed in
the same units as σ (g/cm^2, for example).

   Some materials, notably rubberlike polymers, show elastic
recovery from extensions above the point A' in Figure 2.2. Hence,
in general the elastic limit OB will be above the proportional
limit OA. In many materials (though not all polymers), there is
a stress slightly greater than the elastic limit at which a very
small increase in stress results in a large increase in extension.
The stress OC at which this occurs is called the <u>yield stress</u> or
<u>yield point.</u> In metals the increase in extension at this point may
even by accompanied by a drop in stress.

   In not all textile materials is the Hookean region OA' as prom-
inent, nor are the points A', B', and C' on the stress-extension
curve as distinct, as they appear in Figure 2.2. For many fiber
types the curve departs from linearity very early in the test, so
that it is difficult to identify a proportional limit. Some materials,
such as certain nylons and silk, exhibit no yield point before rupture. In metal technology the points B' and C' are considered to
coincide; while for steel, in particular, all three points A', B',
and C' coincide. Not shown on the curve in Figure 2.2 is a segment frequently found in actual chart tracings for textile fibers.
Such a segment would practically coincide with the negative extension axis. This phase of the extension of the specimen corresponds to the "unbending" associated with removal of crimp in
the specimen. Its behavior in this region is a function of the
configuration imposed on the specimen, and not of the intrinsic
character of the material.

   In textile technology, where tenacity (rather than stress) is
taken as the property responsible for resistance to deformation,
the proportionality constant corresponding to Young's modulus has
been called <u>elastic modulus</u>, <u>elastic stiffness</u> [Reference 90, pp. 17
and 47], and more recently, <u>textile modulus</u>.[2] As Young's modulus
is expressed in the same units as stress, so the textile modulus is
expressed in the same units as tenacity (g/den or g/tex, for example).*

____

   * For a fuller discussion of the terminology used in the regular
engineering and the textile systems, for various mechanical prop-
perties, the reader is referred to Appendix C of Smith's Marburg
Lecture.[90]

The tensile measurements from which the stress-extension curves are obtained are usually made under quasi-static conditions; that is, with application of the force at such a slow rate that it acts like a dead load. Under these conditions the proportional limit of a Hookean material will be found to have a specific value.

2.3 Behavior Above the Elastic Limit

2.3.1 Plasticity. Many materials, as for instance, steel, exhibit perfect elasticity at the levels of load (below the elastic limit) at which they are normally used. If loaded beyond the elastic limit of its constituent materials, however, a body will develop a permanent set, a deformation from which it cannot recover. The mechanism of internal rearrangement by which this occurs is termed plastic flow. In general, a body deformed plastically will recover some of its original size and shape. Thus, if a rod is loaded to some point D' in Figure 2.2, above the elastic limit, and is then immediately unloaded, the stress-extension relationship would have the general character of the dashed line D' D. The unloaded bar would be permanently extended by the amount OD. The degree of recovery varies for different materials; in most metals and textile fibers the slope of the recovery curve is more or less like that of the line D'D, while in lead or putty the recovery curve is practically vertical.

Another important feature of the behavior of solids in the region of plasticity, besides permanent set, is that (in practically all materials) under continuous load beyond the yield point the deformation is time-dependent. This is to say that if a body is loaded to the point D', and the stress is maintained at the corresponding value, the extension will increase indefinitely. In general, some of the extension will be recoverable, as it is when the body is unloaded immediately from the point D', but most of the additional deformation under constant stress will be permanent. This phenomenon of progressive deformation under constant load is known as creep.

The plasticity of solids, as reflected in the tendency for creep to occur, bears a similarity to the viscous flow of liquids. The application of force (gravitation or other forms) is required for the flow of all real (viscous) liquids. Viscous materials are distinguished from solids in their elastic region in that, in the former, there is no unique stress corresponding to each level of strain. Stress arises from deformation, to be sure, but its value depends on the rate of deformation, becoming zero when this rate becomes zero (when the deforming process ceases). Thus, in purely viscous flow, there is no force tending to restore the body to its original size and shape. Any force, however small, produces an irreversible deformation that continues to increase for as long as an external force is applied. Plastic flow, in its general nature, is the same as viscous flow except that, as indicated in a preceding paragraph, the applied force or the stress must reach a yield value greater than zero, before any substantial flow starts.

2.3.2 Viscoelasticity. Among the phenomena considered in the foregoing sections are the two extreme situations: a solid behaving elastically, with no permanent deformation resulting; and a viscous liquid flowing, with no spontaneous recovery possible. These different types of behavior are considered, in the ideal cases, to define distinct forms of matter, or at least, to define sequential stages in the deformation of a given sample. In the real cases of textile and many other polymeric materials, however, the mechanical behavior reflects simultaneous elastic and viscous effects. In most cases, no fixed elastic limit exists; flow occurs along with elastic deformation, and the total deformation at any instant is the sum of those due to the two mechanisms. Recovery from the loaded condition may involve only the elastic deformation. In other cases, the elastic recovery may be quite complete, but be retarded because of internal viscous resistance. When a body of such material is loaded very rapidly, the overall strain will lag behind the applied force, so that, in general, the proportional limit may be changed. Or if a rapid cycle of loading and unloading is imposed on the specimen, the linear segment of the stress-extension curve may disappear altogether and be replaced by a loop.* The property of matter giving rise to these types of behavior is called viscoelasticity. It is exhibited in varying degrees by all rubbers and plastics, as well as by textile materials.

A better insight into the nature of viscoelastic behavior may be obtained from a brief consideration of the molecular mechanism of deformation. The conventional, natural, and man-made textile materials consist of aggregations of long, chainlike molecules. In general, these are supposed to have, in an unstrained fiber, a variety of configurations ranging from a form in which the molecule is curled up on itself to a nearly-straightened form. Few molecules will be straightened to their full, maximum length. The molecules are thought to be intertwined with each other in varying degrees, and to have some chemical forces acting between them. In most textile fibers, the chain molecules tend to be lined

---

* In the latter case, the stress is no longer a single-valued function of extension; and while there is at each point of the curve a rate of change of stress with extension (varying from point to point), the material cannot, strictly, be considered to have a modulus. The existence of a modulus implies a constant rate of such change, the modulus being that rate.

up along the fiber axis. The situation is shown schematically in Figure 2.3. Shown within the dotted lines are crystalline domains, usually called crystallites, in which, over limited portions of their lengths, a number of molecules are locked in an orderly arrangement with respect to each other. Modern theory, which has come to the fore during the last few years (early 1960's), suggests that a single chain molecule may fold back on itself many times within a particular crystallite, before continuing on, out of the crystallite, into an adjacent amorphous region. A large portion of the ordered linear elements in a crystallite may thus be segments of a single chain molecule.

Fig. 2.3. Configurations of chain molecules in a typical, unstrained textile fiber.

A tensile force applied to a fiber tends to straighten out the molecules, and improve their orientation along the fiber axis. In the first stage of this process, some molecules that are already nearly straight will become completely so, and will resist further extension with primary valence forces. In this configurational state, the molecular network may be thought to have reached its elastic limit; for with the removal of the external force the weak stresses, arising from secondary bonds between molecules, would supposedly return the network to its original pattern. With the application of tensile force beyond this limit, the flow of molecules past each other — somewhat like the viscous flow of layers of liquid — sets in. Stresses arise between segments of different molecules; local kinks and bends are removed, and an increasing number of molecules are straightened out. The onset of this flow corresponds to the yield point of the material. With the removal of the external force, the flow in the direction of this force will cease. However, in general, the molecular configurations at that instant will not persist. A new system of secondary-bond stresses (and also, perhaps, of some primary-bond) now exists, and will return the network to a less-extended pattern, though not the original one. This process corresponds to the partial recovery mentioned in connection with Figure 2.2.   If the external force is not removed, the flow will be maintained, and more and more molecules will become oriented. With the improved orientation of the chain molecules will come increasing resistance to the applied force, so that in the later stage of the deformation of the network, equal increments of extension will require increasingly larger force. This situation in the molecular network supposedly accounts for the rise in the stress-extension curve of Figure 2.2, represented by the (sometimes concave-upward) segment D'E'.[*]

2.3.3 Rupture. When the strain at some point in the network (as finally oriented) reaches a certain value, the stress arising from the cohesive force of the valence bonds attains a maximum. With the increase of strain beyond this point, the cohesive force declines. Thus the applied force is no longer balanced by stress at this point, so that more force is available for application at other points in the same cross section.   The same process is repeated there, with the result that very quickly the total force applied to the network ceases to be balanced by stresses in the network.   With greatly reduced resisting stresses acting through the cross section under consideration, the portion of the network on the side of the applied force is rapidly accelerated in the direction of that force. This mechanism is, in brief, the molecular aspect of the process by which the observed rupture, or breakage, of the specimen occurs.

In a tensile experiment, the breaking point is characterized by the breaking stress, the maximum stress that is attained as rupture occurs, and the breaking extension, the corresponding strain. Breaking stress is frequently called the ultimate tensile strength. In this specific sense, then, tensile strength has the dimensions of stress, and is properly expressed in such units as g/cm^2 or $lb/in.^2$. In textile terminology, since stress is replaced by tenacity, the breaking point is characterized by the breaking tenacity, defined analogously to the breaking stress.   Breaking tenacity, evidently, is expressed in the same units as tenacity and textile modulus: g/den or g/tex.

---

[*] A detailed discussion of aspects of viscoelasticity that are outside the scope of the present work is given by Alfrey[1] and, more recently, by Ferry.[30]

2.3.4 <u>Work or Energy of Deformation.</u>   A straining force, in acting on a body through a finite distance, does work on that body. The amount of this work, or expended energy, in a tensile experiment, depends upon the force acting at each instant and the distance through which the end of the specimen moves during the stretching. A mathematical discussion of the subject, involving the calculus, and providing more exact definitions, is given in Chapter 4. Suffice to say here that the <u>strain energy</u> (as it is called) at any extension is proportional to the area under the stress-extension curve up to that point. Expressed in terms of the coordinates of such a stress-extension graph as that of Figure 2.2,* the area under the curve is strain energy <u>per unit volume</u> of material in the specimen.

If a body is strained to some point below the elastic limit, and the load is then removed, the stored strain energy will be recovered. This is to say that the body can do work on its environment in recovery. However, if the body is loaded beyond the yield point, so that plastic or viscoelastic flow occurs, energy expended in producing this flow is lost, being dissipated as heat. Thus, on removal of the load, only a part of the energy expended in the deformation will be recovered. The ratio of the recoverable work or energy to the total expended (area DD'G/area OC'D'G in Figure 2.2) is known as <u>resilience</u> or <u>degree of resilience</u> [Reference 90(p. 50)]. If the body is strained to rupture, no recovery is considered to occur, even though in some cases the broken ends may partially retract. The energy required to achieve rupture is variously denoted <u>work to rupture</u> or <u>breaking energy</u>. If E' is taken to be the breaking point in Figure 2.2, the area OC'D'E'F will be the breaking energy per unit volume of material.

When breaking tenacity is used, instead of breaking stress, to characterize the force agency associated with rupture, the energy involved in the process is thereby referred to some <u>unit of mass</u> (rather than unit volume) of the materials in the specimen. This work or energy parameter is called <u>breaking toughness</u> [References 2,90 (p. 49)]. If the breaking tenacity is expressed in g/den the breaking toughness, as given by the area under the tenacity-extension curve, will be g-cm/den-cm, or g/den. Analogously, the breaking toughness may be given in g-cm/tex-cm or g/tex.† A quantity derived from physics and engineering usage, equivalent in fundamental dimensions to breaking toughness, is <u>breaking energy density.</u> This parameter, employed by Smith, Schiefer, and colleagues, is discussed in Chapter 4.

---

*In performing this computation it should be borne in mind that percentage extensions are to be expressed as fractions.

† The reader is cautioned that when computed from data in these textile units, the breaking toughness is proportional, but not equal, to the energy per unit of mass in the C.G.S. system, or energy/gram

Chapter 3

METHODS AND APPARATUS FOR IMPACT TESTING

3.1 Introduction

The methods of impact testing of textiles are somewhat influenced by the end use to which the particular structure is to be put. While the behavior of a material under certain conditions may be inferred to some extent from properties determined under different conditions, it is preferable, in the interests of the reliability with which the results can be applied, that the conditions under which measurements are made duplicate as nearly as possible those that will be encountered in service. Thus, for example, the ability of a static-line webbing to withstand the shock loading to which it might be subjected would be best evaluated in an experiment in which the impact is made longitudinally. On the other hand, the effectiveness with which a fabric destined for use in personnel armor will defeat free-flying missiles is measured in a test in which the specimen of fabric is impacted transversely by a projectile, at ballistic velocities.

Impact testing apparatus may be classified according to the method used to achieve the rapid loading. Gravity devices, such as the pendulum and falling weight, have been used on textiles for fifty years (see Chapter 1). The pendulum-type testers depend essentially on a falling weight to apply the impact load to the test specimen. However, their design and operation differ enough from other falling-weight types that it is logical to consider the two as separate classes. To achieve higher impact speeds than can be attained conveniently by relying on the acceleration of gravity, the rotating-disc machines and free-flying, ballistic missiles have been more recently adapted to the testing of textiles. Finally, in the last few years, have come pneumatic and hydraulic devices that have been used for the impact loading of textiles, among other materials.

3.2 Falling-Pendulum Impact Tester

3.2.1 Theory.
The mechanical principles of operation of this type of impact tester are those of the compound pendulum. The pendulum proper consists of a bar suspended at one end from a pivot point, about which it is free to rotate in one plane, and having at the other end a bob, in which most of the mass of the assembly is concentrated. The essential features are shown in

21

in Figure 3.1. In operation, the bar is displaced through some angle θ_0 from its freely hanging, vertical position (usually 90° or less) and held there. On release, the bar and mass swing downward under the force of gravity. At the lowest point in its path the pendulum applies the impact load. (The alternative experimental arrangements by which this is accomplished will be discussed in a later section.) After breaking the specimen, the pendulum continues its swing and rises to an angle θ, which is less than θ_0. In practice, θ_0 is fixed, in a given experiment, while θ varies from specimen to specimen. The angle θ is indicated on a dial at the axle of the pendulum, or a circular scale located near the path of the bob; either dial or scale is centered on the axis of suspension of the pendulum.   From the angular displacement and the parameters of the apparatus, without auxiliary equipment, the energy absorbed by (or work done upon) a textile specimen in breaking or tearing it can be calculated.

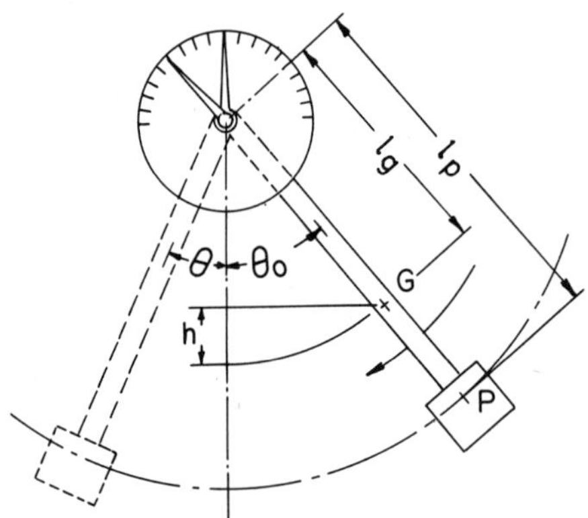

Fig. 3.1. Schematic diagram of falling pendulum, showing geometric quantities involved in measurements with the instrument.

In Figure 3.1, G is the center of gravity of the combined mass M of all the moving parts of the pendulum, ℓ_g is the distance from the center of suspension (pivot point) to the center of gravity, and h is the height of the center of gravity G before release of the pendulum above the position of G at the instant of impact.

The energy available for extending or rupturing the specimen is the kinetic energy of the pendulum at the point of impact. This energy is given by the potential energy V_0 of the pendulum in its raised, stationary position, less the energy lost in friction in the bearings and in air resistance in the first downward swing of the arm. While the frictional losses are generally small, and not readily measurable directly, they may be eliminated from the calculations by consideration of the potential energy V_1 of the pendulum at the end of a full swing with no specimen in the apparatus. This is evidently

$$V_1 = V_0 - H_1 \tag{3.1}$$

where H_1 is the frictional energy lost in the full swing.

Now V_0 is given by the familiar expression Mgh, g being the acceleration of gravity. In turn, $h = \ell_g(1 - \cos\theta_0)$, so that

$$V_0 = Mg\ell_g(1 - \cos\theta_0) \tag{3.2}$$

Similarly,

$$V_1 = Mg\ell_g(1 - \cos\theta_1) \tag{3.3}$$

where θ_1 is the angle of deflection of the bar at the end of its "dummy" swing. Clearly, then, the energy lost in friction is

$$H_1 = V_0 - V_1 = Mg\ell_g(\cos\theta_1 - \cos\theta_0) \tag{3.4}$$

If the experiment is now performed with a specimen in place, the pendulum at the end of its swing will have a potential energy $V < V_1$. In practice, V_0 is fixed, in a given experiment, while V varies from specimen to specimen, because of nonuniformity or sample differences. The loss in energy $V_0 - V$ will include the energy A_d expended in deforming, breaking, or tearing the specimen, as well as a frictional loss H, distinct from H_1. As in the case of Equation 3.4,

$$A_d + H = V_0 - V = Mg\ell_g(\cos\theta - \cos\theta_0) \tag{3.5}$$

where θ is the angle of deflection of the bar at the end of its swing with the specimen in place. We may reasonably assume that H_1 and H are proportional to the distances of travel of the pendulum arm in the two cases. Hence,

$$H = \frac{\theta + \theta_0}{\theta_1 + \theta_0} H_1 \tag{3.6}$$

$$= \frac{\theta + \theta_0}{\theta_1 + \theta_0} Mg\ell_g(\cos\theta_1 - \cos\theta_0) \tag{3.7}$$

Subtracting Equation 3.7 from Equation 3.5, we obtain for the strain or rupture energy

$$A_d = Mg\ell_g\left[\cos\theta - \left(1 - \frac{\theta + \theta_0}{\theta_1 + \theta_0}\right)\cos\theta_0 - \frac{\theta + \theta_0}{\theta_1 + \theta_0}\cos\theta_1\right] \tag{3.8}$$

In practice, the mass M is usually adjusted for a particular sample so that θ does not differ greatly from θ_1, thus making the frictional losses nearly equal in the two cases. When this is done, the term $(\theta + \theta_0)/(\theta_1 + \theta_0)$ becomes nearly equal to unity, and $1 - (\theta + \theta_0)/(\theta_1 + \theta_0)$ becomes negligibly small. Furthermore, the latter term is the coefficient of a small fraction ($\theta_0 = 90°$ approx), so that the product may be neglected. Thus, Equation 3.8 reduces to the approximate form,

$$A_d = Mg\ell_g(\cos \theta - \cos \theta_1) \tag{3.9}$$

It is by means of this equation that the work done upon the sample, in deforming or rupturing it, is calculated.

The masses of the pendulum should be so distributed with respect to its linear dimensions that the point of impact is at the center of percussion. When this is so, the pendulum, on impacting the specimen, will tend naturally to continue its rotation about its axis of suspension. If the impact occurs at any other point, the pendulum will tend to rotate about an axis different from that of the shaft on which it is suspended; the bearings will experience a jar, and vibrations will be set up in the pendulum bar and its supports. Thus, some of the kinetic energy at the instant of impact will be dissipated in the internal friction associated with these vibrations. This loss being indistinguishable from the work done on the specimen, the quantity measured, as work, tends to be fictitiously high.

The kinetic energy of the falling pendulum at the instant of impact is equal to the initial potential energy Mgh, less the frictional loss incurred in the falling of the pendulum through the angle θ_0. Since, as has been implied earlier in this section, the latter loss is negligible, we have for the angular kinetic energy on impact

$$\tfrac{1}{2}MK^2\left(\frac{v_0}{\ell_p}\right)^2 = Mgh \tag{3.10}$$

where K is the radius of gyration of the pendulum, v_0 is the velocity of the center of percussion (impact velocity), and ℓ_p is the radius of percussion. The radius K is given by the formula

$$K^2 = \ell_g\ell_p \tag{3.11}$$

Hence, Equation 3.10 may be simplified,

$$\left(\frac{\ell_g}{\ell_p}\right)v_0^2 = 2gh$$

so that the impact velocity

$$v_0 = \sqrt{\frac{2gh\ell_g}{\ell_p}} \tag{3.12}$$

The radius of percussion, the distance ℓ_p (Figure 3.1) from the suspension axis to the center of percussion, is given by the relation

$$\ell_p = \frac{I}{M\ell_g}$$ (3.13)

where I is the moment of inertia of the moving parts of the pendulum. The moment I may be calculated from the masses and dimensions of the various parts of the pendulum, according to standard formulas. If the geometric forms of the parts of the pendulum are simple, which is frequently the case, such a calculation is not very difficult. A more convenient method for locating the center of percussion depends on the period T of the pendulum, that is, the time required for one full swing from any position back to that position again. If means are available for measuring T accurately, this method is perhaps preferable. The radius of percussion is then given by

$$\ell_p = \frac{gT^2}{4\pi^2}$$ (3.14)

3.2.2 Methods of Mounting the Test Specimen. There are, basically, two ways in which the impact load can be applied to the specimen; either longitudinally, or transversely. By far, most of the adaptations of the falling pendulum tester to textiles have employed the former type of loading. In the method introduced by Lester[47] and Midgley and Peirce[66,76] for a breaking test, the specimen is attached at one end to the bob of the pendulum, and at the other end to a fixed, rigid support, with suitable clamps, as shown in Figure 3.2. If the specimen is long enough, it may

Fig. 3.2. Falling-pendulum impact tester used by Midgely and Peirce.[66]

be clamped directly to the bob and the fixed support.   Adjustments are made so that the specimen is taut when the pendulum is at the lowest point in its swing.

For cases where the specimens are too short to be mounted in this fashion, another method is to make the specimen only a part of the line from the pendulum to the fixed post. The connecting element should be flexible enough that it does not impede the motion of the pendulum, and should have an extensibility that is negligible in comparison with that of the specimen, to avoid an unknown amount of energy being lost in stretching this element. Balls,[10] and later, Lang[45] have described such arrangements for breaking cotton, wool, and other staple fibers.

Lang's description of his method is as follows: "The sample was mounted in identically shaped clamps, the leading one being hooked on the rear of the pendulum and the following clamp being fixed to a length of brass chain attached at a suitable height behind the pendulum, to a rigid anchorage on the base of the instrument. As the pendulum fell, the chain followed, until the bottom of the swing was reached, where, when it became taut, the sample broke and the chain fell to the base. The leading clamp travelled with the instrument. The length of the sample and the position of break were easily adjustable.   The speed and efficiency of operation were considered satisfactory." (Views of the device are shown in Figures 3.3 and 3.4).   While this method was considered successful, inaccuracies were introduced by the tendency of the chain to flip upward when the specimen broke. This action would result in a loss of energy that would be reflected in the reading obtained on the instrument.

Other methods for breaking staple-fiber bundles by impact were devised by Lang, but only one of these seems to have been acceptable. This one is based in principle on the modification at the Shirley Institute of the pendulum instrument of Midgley and Peirce, mentioned in Chapter 1. According to Lang, in his apparatus, "the pendulum bob of cylindrical shape, with its long axis perpendicular to the plane of motion and suspended by two rods at its ends, was slotted through its lower half in the center, so that it might swing astride a narrow, adjustable pedestal on the base. To provide a flat contact surface for the foremost clamp, which was fitted with a stout crossbar, short lateral slots were made on either side of the front face of the above-mentioned slot at the level of the center of percussion of the system. Two small brass clips or, preferably, small adhesions of 'Plasticine' on the crossbar arms enabled the pendulum to retain the leading grip, after breaking the sample." (A sketch of the arrangement is shown in Figure 3.5.)

In the much larger Shirley instrument, the forward clamp is suspended by means of two rods, from the same shaft as the

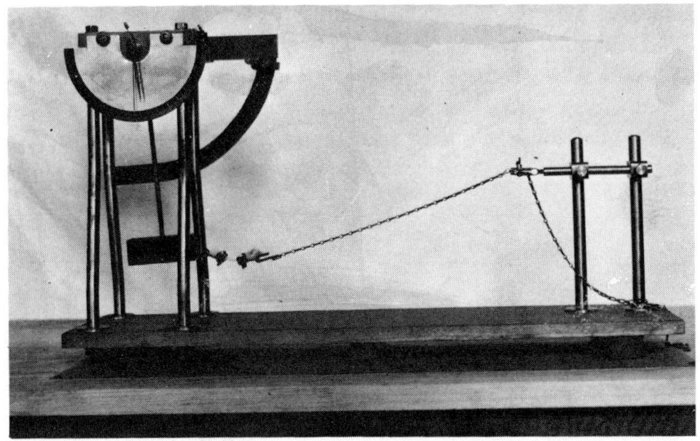

Fig. 3.3. Pendulum tester using trailing chain; specimen about to be broken.[45]

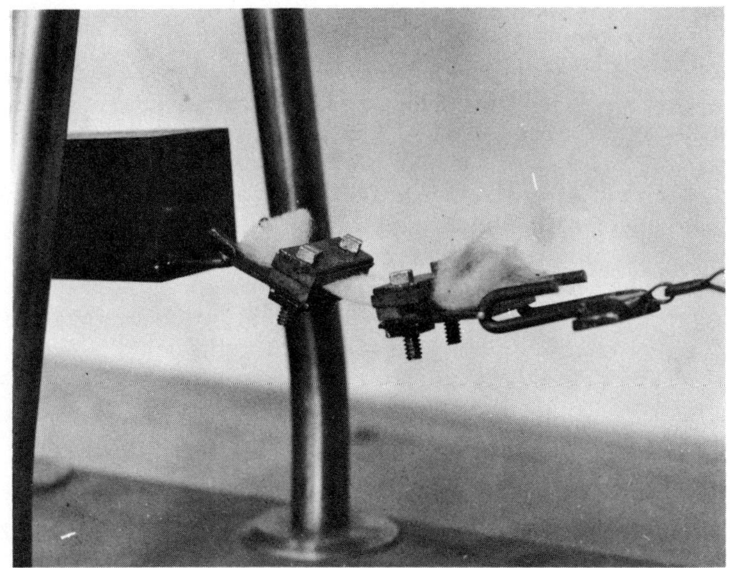

Fig. 3.4. Pendulum tester using trailing chain; close-up view of specimen and clamps.[45]

main pendulum, thus forming, as has been noted in Chapter 1,
a second pendulum.[15] The rear end of the test specimen is attached to a rigid post on the base of the machine. Corresponding
to the crossbar in Figure 3.5 are means of engagement with the
falling pendulum at the center of percussion, in its lowermost
position. The forward clamp is carried along by the impact,
thus deforming and (in most tests) separating the specimen.

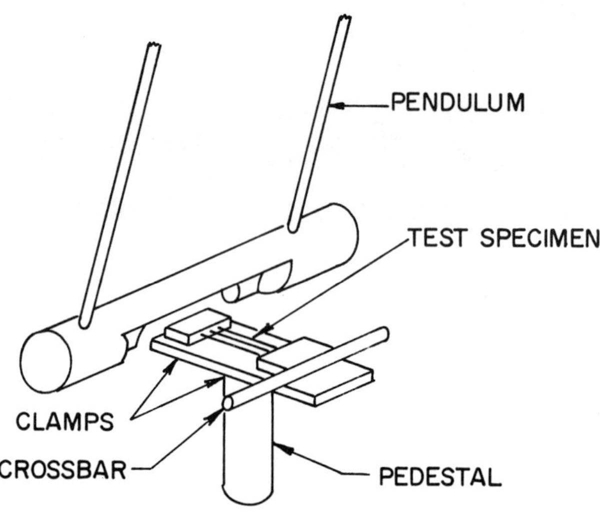

Fig. 3.5. Pendulum tester using stationary crossbar and
clamp; close-up view including bob. [After Reference 45.]

Some of the kinetic energy of the falling heavy pendulum goes
into setting the forward-clamp assembly in motion (in the Lang,
as well as in the Shirley apparatus).   However, Equation 3.9
may still be used to determine the energy expended on the specimen, if θ_1 is determined with the forward clamp in its engagement position. Although the apparatus is described as being used
for impact tearing tests on fabrics, it is equally capable of tensile breaking tests on heavy yarns and cords, and fabrics.

Another method for mounting yarns and cords for longitudinal
impact tests was devised by Lyons and Prettyman.[51]   This has
been described by them as follows: "Mounted in supporting
brackets on the right-hand edges of the supporting frame F [see
Figure 3.6] at about the level of the center of percussion of the
pendulum in its equilibrium position, are the cord clamps E, of
the conventional snubbing type. Opposite the clamps, on the left-
hand side of the instrument, is a light extension member G, to
the outer end of which is affixed, by means of binding posts, the
cord K, used for aligning the test specimen L, in the path of the
pendulum." The scheme for mounting and breaking the test specimen is clearly depicted in Figure 3.6b.

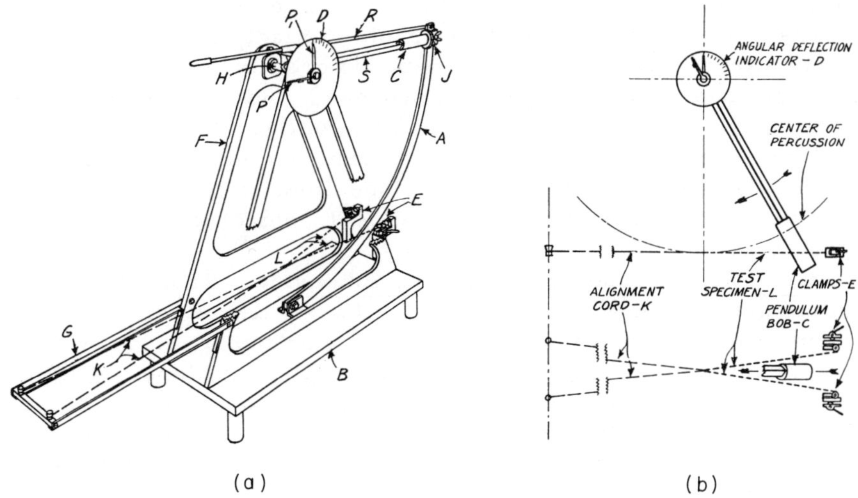

Fig. 3.6. Pendulum apparatus for impact rupture tests on textiles, principally tirecords.[51] (a) Over-all, perspective view of instrument. (b) Schematic drawing, showing position of test specimen, with reference to pendulum proper.

While this method of mounting places the specimen in the path of the pendulum, the loading of the specimen is essentially longitudinal, rather than transverse. It was noted that "in practically no case has rupture been found to occur in the segment of the test specimen with which the bob comes into contact. The breaks generally occur in one or the other of the two straight sections of the looped specimen." Acceptably reproducible results were obtained with this method of mounting; coefficients of variation in the range 2.6 to 7.7% being found for cord toughness values for rayon and nylon tirecords.

Contrariwise, the mounting of specimens at the bottom of the pendulum swing, so that the impact loading of the specimen is initially, and remains, predominantly transverse, seems consistently to lead to great variability. Morton and Turner,[71] in impact tests on fabrics, mounted their specimens in a vertical plane across the path of the pendulum. They found qualitative, as well as quantitative, variability in their results, noting that some specimens gave "values as much as 70% greater than others from the same piece of cloth." Such results seem largely to account for these workers' dissatisfaction with the impact test. Lyons and Prettyman made some initial trials with single-cord specimens mounted perpendicular to the plane of the pendulum, in the path of the bob. Independently, they reached the same conclusion as Morton and Turner, stating that "with this method of mounting, the observed individual values were found

to be too scattered to be acceptable."[51] The deficiency in transverse mounting evidently arises from the fact that the resisting
force exerted by the specimen depends greatly on its varying
configuration during the impact process. This follows from the
law for the resolution of forces. The configuration and the pre-
vailing tension in the specimen, in turn, are highly sensitive to
the initial tension under which the specimen is mounted. This
initial tension evidently requires closer control than has been
practicable.

3.2.3 Measurement of Force and Extension. From the foregoing discussion of the theory of the falling pendulum, it is evident that only the energy of deformation or rupture, and the velocity
of impact can be measured with this instrument alone. Of fre-
quent interest, however, is the force-extension curve of a sample
under impact conditions. To obtain information of this nature,
auxiliary apparatus must be employed in conjunction with the
pendulum.

For such supplementary apparatus, Meredith[62] used a cantilever spring for measuring tension in the specimen, and a rotating-
drum camera to obtain the tension as a function of time.   The
yarn specimen was mounted between a clamp attached to the pendulum at its center of percussion, and another clamp on the vertical cantilever spring. The geometry of the system was such that
the impact load was applied to the yarn at the bottom of the swing
of the pendulum, thereby deflecting the free end of the spring. A
mirror on the spring provided means for reflecting a beam of
light to photographic paper in the camera.[*]

Of these innovations, Meredith says, "The movement of the
tension-measuring spring was arranged to be very small (0.003
cm. for tension of 1,000 gm.) in order to reduce inertia errors
and to enable a constant rate of extension to be attained.   This
movement was magnified (790 ×) by an optical-lever system so
that the calibration factor was 425 gm./cm. deflection on the recording paper. . . .

"A synchronous motor geared to give 25 r.p.s. turned the 2-
inch diameter drum which carried the recording paper.   The
record obtained from the instrument was a load-time record from
which the breaking load and time to break could be deduced directly. If the velocity of the pendulum could be considered constant during the extension of the specimen, then the record would
also be a load-extension curve.

"This constancy of velocity was achieved by making the kinetic
energy of the pendulum large compared with the energy required

---

[*] This method for measuring and recording tension in the specimen is similar to that later used by Meredith in a rotating-disc
instrument, as shown in Figure 3.10.

to rupture a specimen and ensuring that the extension of yarn took place with the pendulum at the bottom of its swing so that its kinetic energy was changing very little. . . ." The constancy of the velocity, of course, was only approximate since some energy was absorbed during the rupture process, and the pendulum therefore retarded. Meredith cites as an example, however, an experiment in which the initial impact velocity was 45 cm/sec, while the calculated average velocity was 44 cm/sec, a difference of only 2%.

The breaking elongation was calculated by multiplying the mean time to rupture, obtained from the photographic record, by the average velocity of the clamp at the center of percussion.

High-speed photography of the moving pendulum during the impact process was used by Lyons and Prettyman[51] to derive force-extension curves of tirecords. They set up a scale at the level of the bob as it approached, broke, and left the specimen held in its path. A pointer attached to the bob was so positioned as to traverse the face of the scale. A high-speed camera was set up to view the scale at such a distance that the scale nearly filled a frame on the film. By operation of the camera during the impact process, a photographic record of the position of the pointer on the bob at fractional-millisecond intervals was obtainable.

These writers go on to describe this technique further: "By means of a microfilm reader, scale-and-pointer readings were taken from the film, frame by frame. The readings were then plotted on a frame-_vs._-distance graph. Both film records gave, for the initial stage, a linear relationship between time (as measured by number of frames) and distance, indicating that the pendulum had not yet reached the cord. This section of the graph was followed by one of downward curvature, representing the deceleration of the pendulum as it broke the cord. The final section of the graph was again linear, as the pendulum bob moved away from the broken cord.

"An enlarged plot of the curved section of each graph was next made, and on this, slopes at every 2.5 frames were read. The results of these readings were, in turn, plotted on a frame-_vs._-velocity (in./frame) graph, and a smooth curve of best fit was drawn through the plotted points. Finally, slopes were read on this graph, giving a relationship between deceleration and frames. The deceleration, at this stage, was conveniently expressed in 0.0005 in./frame^2. Instead of plotting the deceleration data against frame numbers, however, recourse was taken to the initial plots relating distance with frame number. Thus, it was possible to plot deceleration as a function of distance of pendulum travel (or cord elongation).

"Using the known speeds of the films, the [units in which decelerations were expressed] were readily converted to cm./sec.^{2},

and these, in turn, converted to angular measure by dividing by
the radius of oscillation (93.1 cm.) of the pointer on the pendulum bob. The torque of the breaking cord, exerted on the pendulum so as to decelerate it, is given by the relation

$$T = I_p \alpha \tag{3.15}$$

where I_p is the moment of inertia of the pendulum (11.1×10^6
gm.-cm.^{2}), and α is the angular acceleration (or deceleration).
The decelerating force, or force with which the cord is progres-
sively loaded during the rupture process, is given by

$$F = \frac{T}{d} \tag{3.16}$$

where d is the distance of the impact point on the bob from the
axis of suspension (78.5 cm.). By means of these considerations,
it is possible to interpret the deceleration as loading force, re-
membering that this force is divided between the two halves of
the test cord on either side of the pendulum bob."

Ballou and Roetling[9] have described another method, which they
followed, to obtain the stress-strain curves of textile specimens
under impact conditions, using a falling pendulum. A schematic
diagram of their apparatus appears in Figure 3.7.   They state:
"The specimen to be tested is mounted with one end attached to
a transducer and the opposite end attached to a clamp (referred
to as the head clamp) which rests upon a support (support not illustrated) but is otherwise free to move. This clamp is picked
up by a falling pendulum and moves with the velocity of the pendulum to strain the yarn. The stress-strain curve is obtained
by connecting the amplified output signal of the transducer to the
vertical deflection plates of an oscilloscope while at the same

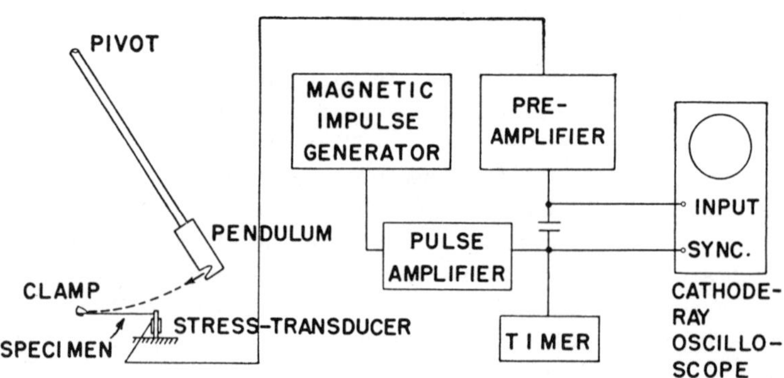

Fig. 3.7. Schematic diagram of pendulum apparatus using a
stress transducer, with auxiliary electronic instruments.[9]

time applying a signal proportional to strain to the horizontal plates. Thus the stress-strain curve is displayed on the face of the oscilloscope and is recorded by means of a 35-mm. camera." A similar method was used by Parker and Kemic[31] for impact studies on tirecords. The latter employed a ring potentiometer connected to the horizontal-sweep plates of an oscilloscope for strain measurement.[83]

Theoretically, of course, there is no upper limit to the impact velocity attainable with a falling-pendulum tester. Higher and higher impact velocities evidently can be attained by increasing the radius of percussion, as well as by raising, toward 180°, the angle through which the pendulum is made to fall before impact. In practice, however, considerations of convenience of operation and economy of floor space have limited the size of the machines, and hence the available maximum impact velocity. Typical of the order of impact velocities attainable in actual pendulum devices are the 3.8 m/sec of Lyons and Prettyman, and 2.1 m/sec of Ballou and Roetling. The latter obtained a high rate of extension, 10^5 %/min, by using a short (5 in.) gage length. Impact velocities in a higher range are attainable with falling-weight types of testers, which are more compact than falling pendulums giving comparable velocities.

3.3 Falling-Weight Apparatus

3.3.1 Theory. Falling-weight impact testers are similar to pendulum testers in that the velocity or energy with which the load is applied to a specimen depends on the height from which a weight is allowed to fall before engaging the specimen. While in the pendulum tester, the weight is constrained to follow a circular path, so that its velocity is horizontal at the instant of impact; in a falling-weight apparatus, a mass is allowed to fall freely, and follows a vertical path.

In falling-weight devices, the velocity of impact is that of the center of gravity of the weight. This velocity is given by the equation:

$$v_g = \sqrt{2gh} \tag{3.16a}$$

where h is the distance through which the center of gravity falls. Correspondingly, the kinetic energy of impact is given by the expression Mgh, where M, in this case, is the mass of the falling weight. This expression for the impact energy is based on the reasonable assumption that no energy is dissipated in friction during the descent of the weight, before engaging the specimen.

Leaderman[46] has described two methods in which the falling-weight tester may be used. In the first of these, which he calls the "variable-velocity-of-deformation" type, the weight is allowed to drop from a height h above the impact bar or weight pan.

The "energy required to break the specimen is obtained by measuring the kinetic energy remaining in the weight after the instant of rupture," the rupture energy being the difference between this residual energy and Mgh. In the other, "zero-velocity-of-deformation" method, the height h "to which the weight must be raised in order just to break" the sample is determined, presumably on a succession of specimens, assumed to be identical. Each specimen would be tested at a different value of h. In this method, the rupture energy is the impact energy, and is given by Mgh, where h now has the limiting value determined by this procedure Since, theoretically, the weight has no residual kinetic energy at the instant of rupture, its velocity, and therefore, the velocity of deformation of the specimen at that instant, is zero.   This method was used by a firm that tested a variety of ropes for du Pont.[29]

Leaderman has commented that this zero-velocity method favors samples having a low extension to rupture, since these are under load a shorter time than a more extensible sample would be. A disadvantage of changing the impact energy by altering the height from which the weight is dropped, is that the impact velocity is changed at the same time. Thus, different rupture energies found by this method, for a number of samples, would not be strictly comparable, since the impact velocities would differ. This shortcoming can be avoided, for the impact energy Mgh can be altered by changing the mass M of the falling weight, and by keeping the height of drop h, and therefore, the impact velocity v_g constant. A method similar to this was followed by Schiefer, Appel, Krasny, and Richey[86] in impact tests on yarns made from a variety of fibers. They tested five specimens of a yarn sample with such a weight and height of drop that the impact energy was insufficient to break any of the specimens. Five specimens were then tested at each of a number of higher energies, obtained by adding plates, in equal increments, to the falling weight. The height of drop was held constant, so as to give an impact velocity of about 6000 in./min. The percentage of specimens that failed at each energy level was plotted as a function of impact energy. It was found that a straight line could be fitted to the plotted points. The average rupture energy was then taken as the point on the graph corresponding to the failure of 50% of the specimens in a group.

While the rupture energy was of secondary interest to them, Newman and Wheeler[73] followed this procedure of varying the weight by the addition of plates, keeping the height of drop constant, in order to obtain relationships between impact energy and stretch in nylon and sisal ropes.

<u>3.3.2 General Types.</u> Representative designs of falling-weight apparatus are shown schematically in Figure 3.8. In Figure 3.8a

appears the arrangement described by Leaderman[46] for measurements on heavy yarns and cords. Either of two falling weights
W, having masses of 500 and 25 lb, was used, depending on the
impact energies desired. These had a square C shape in horizontal cross section. The specimen S was hung from a rigid
support R, and attached at its lower end to an impact bar B,
seen end-on in Figure 3.8a. The opening in either weight was
of sufficient size that when the weight was dropped, it would clear
the support R and the clamps holding the specimen. The bar B,
however, was made sufficiently long to be engaged by the falling
weight, thus applying an impact load to the specimen.

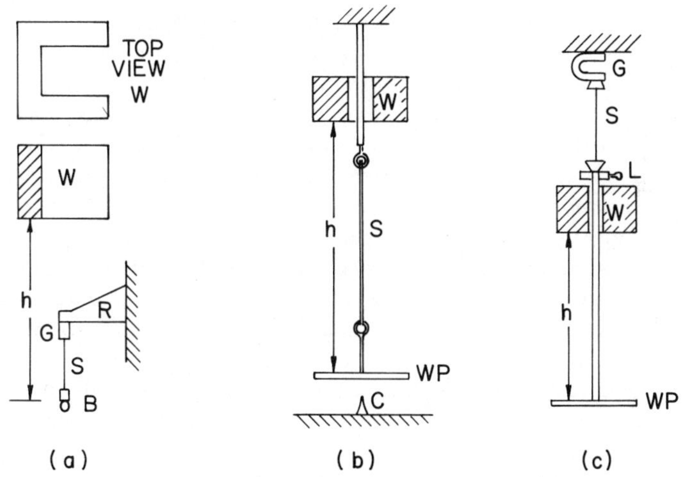

Fig. 3.8. Schematic diagrams of general types of falling-
weight testers used for impact tests.

The method depicted in Figure 3.8b is in principle, that used
by Newman and colleagues,[73,103] by Schiefer and colleagues, and
in the work done for the du Pont Company, cited in an earlier
section. The specimen S is supported, from overhead, by a
massive, rigid framework. The impacting weight W has an axial, vertical hole of sufficient diameter to clear links and clamps
when the weight is allowed to fall down about the specimen. Suspended from the lower end of the specimen, to which it is firmly
attached, is a weight pan WP, or other means of intercepting the
falling weight. A slight modification of this arrangement, in
which the travel of the weight is entirely below the specimen, as
shown in Figure 3.8c, was used by Baker and Swallow.[8]
The measurement of the kinetic energy remaining in the falling weight after rupture of a specimen (outlined by Leaderman,
as mentioned earlier in this section) is not as simple as the cor-

responding measurement in the falling-pendulum type of tester.
A disadvantage of the falling-weight type is that the residual energy is not conserved at the end of an impact breaking test. The
residual energy is dissipated in stopping the falling weight; this
has, in some cases, involved the use of special shock-absorbing
platforms (as on the M.I.T. impact tester, for instance).   The
rupture energy, where it has been obtained, has usually been
calculated as a by-product from measurements of force and elongation as functions of time, with auxiliary electronic equipment.

   3.3.3 Measurement of Force and Extension. Most of the falling-
weight devices that have been used on textiles have been provided
with means for these measurements. In the apparatus described
by Leaderman, Figure 3.8a, a resistance-wire strain gage was
attached (in parallel) to the "weight" bar G from which the specimen S was supported. Thus, by virtue of a prior calibration,
the forces in the specimen could be related to the resistance of
the strain gage. Changes in this resistance, in turn, were converted into voltage changes.   These, suitably amplified, could
be registered on an oscilloscope, or recorded on photographic
film.   Thus, an instantaneous record of the force-time relationship in the specimen, during impact, could be obtained. By con-
ducting the tests under suitable conditions (specimen size, height
of drop, and so forth, the rate of stretch could be made substantially constant. This represents a linear relationship between
stretch, or change in length $\Delta \ell$ and time t: $\Delta \ell = v_g t$, where v_g
is the impact velocity. Thus, this method gives not only a force-
time curve, but a force-elongation curve for the specimen during the impact process.

   Newman and Wheeler[73] devised a method for measuring the
stretch in ropes under impact, but had no means for obtaining
forces in the specimens during the test. These workers placed
a column of clay, C in Figure 3.8b, on the floor beneath the
weight pan WP. They describe their measurement: "The distance from the bottom of the weight pan to the floor was measured before test, and the height of the clay column was measured
after the impact load had been applied. The difference between
the two measurements is the stretch of the specimen under the
impact load. . . . Where all of the strands of a specimen parted,
it was impossible to determine stretch under load."

   In a later modification of this equipment,[103] four wire strain
gages were cemented to the surface of the shank of the eyebolt
from which the specimen hung. Each of these gages formed one
of the arms of an electrical (Wheatstone) bridge, so arranged
that the output was proportional to tensile force exerted on the
eyebolt by the specimen as a result of the impact. Suitably amplified and fed into an oscillograph, this output provided a photographic record of the impact force on the specimen as a function

of time,   (the force-time curve, such as was obtained by
Leaderman). The clay-column technique was used to measure
the maximum stretch of the specimen, but this, obviously, is
inadequate for deriving the force-elongation curve.   For this,
the instantaneous elongation as a function of time is needed. This
relationship was calculated from the differential equation of motion of the dropped weight and weight pan after impact:

$$(M + m)\frac{d^2s}{dt^2} = (M + m)g - F - mg \tag{3.17}$$

where s is vertical distance (elongation), t is time, M is the
mass of W, m is the mass of the weight pan, and F is the force
in the specimen resulting from impact, as given by the photographic record. Since F was thus known as a function of time
F(t), Equation 3.17 could be integrated numerically to yield the
elongation as a function of time s(t). By plotting the measured
forces against the calculated elongations at corresponding times,
the force-elongation curve was obtained.

In the work done for du Pont[29] on ropes of various types, the
specimens were similarly suspended from a stress-measuring
load cell, the output of which was registered on a multichannel
oscillograph. The motion of the falling weight, from the instant
of release to the completion of the impact process, drove the
slider of a potentiometer, through a system of pulleys and cables
attached to the weight. The output of the potentiometer, in turn,
was fed into another channel of the oscillograph, so that the position of the falling weight, and hence, the elongation of the specimen was indicated at every instant during the impact process.
From the tracings of force and elongation as functions of time,
it was possible to calculate the energy absorbed by the specimen,
presumably by measuring the area under the force-elongation
curve. Data for heights of drop just sufficient to rupture a sample were used in the calculations, so that the results reported
were breaking energies.

In the Baker and Swallow apparatus[8] the yarn specimen S was
similarly suspended from a force-measuring gage, G in Figure
3.8c. This was a U-gage having on its open ends plates which
formed an electrical condenser. The downward deflection of the
lower arm when a force was applied, altered the capacitance of
the condenser. This alteration, through electronic circuitry,
was registered as an oscilloscope tracing, which was photographed with a drum camera, rotating at constant, known speed.
The gage system was calibrated by incremental static loading. It
was thus possible to obtain a force-time curve when the specimen
was impact loaded, by allowing the weight W to fall on to the pan
WP. For the measurement of elongation, a small spotlight and
battery were affixed to the rod-and-pan assembly just below the

specimen.   Motion of the point of light when the system was
impact-loaded was photographed with a drum camera, thus providing an elongation-time record. As in the procedures described
above, this and the force-time record, along with the length and
linear density of the specimen, gave the data for plotting the
tenacity-extension curve.

When the weight W was dropped from small heights insufficient to break the specimens, Baker and Swallow observed damped
oscillations in the force-time and elongation-time tracings, presumably about the static equilibrium position of the loaded rod-
and-pan system.   In order to obtain the data for the tenacity-
extension curves, the weight was dropped from a sufficient
height to break the specimens on impact.

In later modifications of the M. I. T. falling-weight tester,
originally described by Leaderman[46] and Schwarz[89] a piezoelectric crystal transducer, having the advantages of high electrical
output and high frequency response, was used to obtain higher
sensitivity for the measurement of force in impact tests on
yarns.[55,56] The signal generated by a device of this type is a voltage proportional to the applied force.   Suitably amplified, this
voltage was displayed on an oscilloscope screen and photographed
to give a force-time record. For force measurements on heavier
structures, such as webbings, a system of four resistance-wire
strain gages was used, substantially in the manner of Stang,
Greenspan, and Newman.[103]

Elongation of the specimen was measured by means of a magnetic-tape extensometer. Prerecorded on the tape was a sinusoidal signal of known constant frequency. The tape was inserted
in guides in the extensometer so as to pass over the open end of
a U-shaped electromagnetic coil. One end of the tape was firmly
attached to the impact bar, which was supported by the specimen.
Thus, in impact, the tape was pulled through the extensometer at
the same rate at which the impact bar, and (assuming proper
clamping) the lower end of the specimen, moved downward. By
virtue of the periodic signal on the magnetic tape, its movement
over the coil generated an alternating voltage, the frequency of
which was proportional to the speed of the tape.   This voltage
was amplified, fed into an oscilloscope, and photographically recorded. With the known prerecorded wave lengths on the tape
and the time of oscilloscope sweep, the resulting record per-
mitted the calculation of the velocity of the end of the specimen,
in selected intervals of time during the impact process.   The
displacement of the end of the specimen, and hence the elongation during the impact process, could be obtained from a count
of the number of waves on the record, since the length on the
tape, of each wave, was uniform and known. A typical record
is shown in Figure 3.9

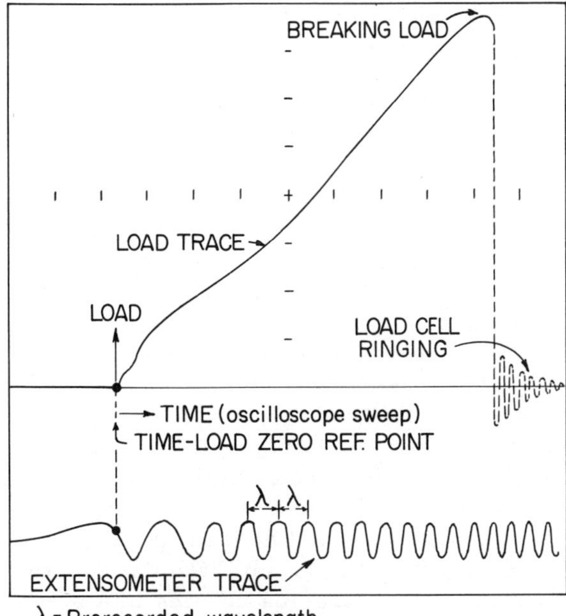

Fig. 3.9. Typical record of oscillograph traces obtained from force transducer and extensometer on M. I. T. falling-weight tester. [After Reference 42.]

Backer and colleagues[7,42] have given another method in which the magnetic-tape transducer has been adapted to elongation measurements. Their description is as follows: "Two or more tapes can be used at one time to describe the displacement history of any designated pair of gage marks, and the differences between these two displacements at any time furnish data on the local strain of that portion of the specimen. Two-point impact strain readings have been found entirely feasible on many textile structures of sheet form. Impact strain measurements on yarn specimens have been taken with a single tape attached to the moving flat jaw in cases where jaw breaks did not occur. This latter tape also provides a check on test velocities. The use of two-point strain reading for impact tests on twisted structures (such as ropes) has been found possible, but the torsional rotation of the rope during the test has interfered significantly with the reliability and reproducibility of the results.

"The strain readings obtained with the magnetic-tape system eliminate the effects of the strain inhomogeneity in the region of the jaws. However, the stress concentrations of such regions may still seriously affect the maximum load readings obtainable for a given textile specimen. Also, to avoid penetration of the strain inhomogeneity into the center of the specimen, it is often necessary to use a specimen of length six to ten times its width." The use of tape in impact testing is limited to velocities below 100 ft/sec.

In reporting on some phases of this later work on the M.I.T. tester, Cheatham[31] noted that because of the elastic aspect of the impact, and the conservation of momentum, the impact bar and lower clamp tended to bounce away from the falling weight. Advantage was taken of this effect, to increase the effective initial velocity. Since the magnetic tape gave a record of the motion of the impact bar, and not of the falling weight, it was possible to measure this initial velocity. It is reported that velocities approaching 40 ft/sec were obtained on yarns, using the 25-lb falling weight, where the velocity of the latter at impact was about 25 ft/sec [Reference 56 (Final Report)].

Another phase of the remodeling of this tester that received important consideration was the installation of safety features in the suspension for the 500-lb falling weight. Originally, the weight was raised by cables and held in its drop position by energized electromagnets. In the modified design, the weight is held by horizontal, solenoid-actuated pins in a movable frame or jacket. Support of the weight is thus of positive nature. The jacket is permanently attached to cables, by which it can be raised to any level up to the full height of the machine. Thus, it is possible to vary the height of drop, and accordingly, the impact velocity, an added feature of the new design.

As noted by Stone, Schiefer, and Fox,[104] the dimensions of laboratory rooms have limited the maximum velocity attainable with pendulums and falling weights to about 20,000 in./min (28 ft/sec), corresponding to a height of drop of about 12 ft. The "bounce" effect may raise this velocity slightly, under certain favorable conditions. To reach appreciably higher impact velocities, recourse has been taken to power-driven devices, such as are described in the following sections of this chapter.

3.4 Rotating-Disc Methods

3.4.1 Background. The rotating disc has been employed in the impact testing of textiles to achieve velocities higher than those attainable with practicable methods dependent on the acceleration of gravity. This device represents the adaptation of the principle of a machine described by Mann[53] twenty-five years ago, for impacting metal specimens. Mann's machine had two parallel, motor-driven discs carrying horns or lugs. These could be released to engage a clamp holding the specimen, when the desired speed had been attained. In the testers of this type that have been built for the testing of textiles in tension, the angular momentum of the disc, or drum, is so great at impact that the loss of velocity or energy, on accelerating, deforming, and

breaking a specimen, is undetectable.   In consequence, with
these devices (unlike the adaptations of the pendulum and, in
some cases, the falling weight) no quantitative information about
the impact behavior of a sample can be obtained without making
observations directly on the specimen.

3.4.2 Method of Meredith. The apparatus used by Meredith,[63]
to obtain constant rates of extension from 20 to 1000 %/sec, in
13-cm specimens of yarns, is shown in Figure 3.10.   Meredith
describes his equipment as follows: "A cast steel cantilever (L)
is mounted on a rigid pillar (P) which is further strengthened by
a steel strut between this pillar and the gear box behind the rotating disk (D) which is driven at different speeds by a synchronous motor (T).   The cantilever carries a grip at its free end, and

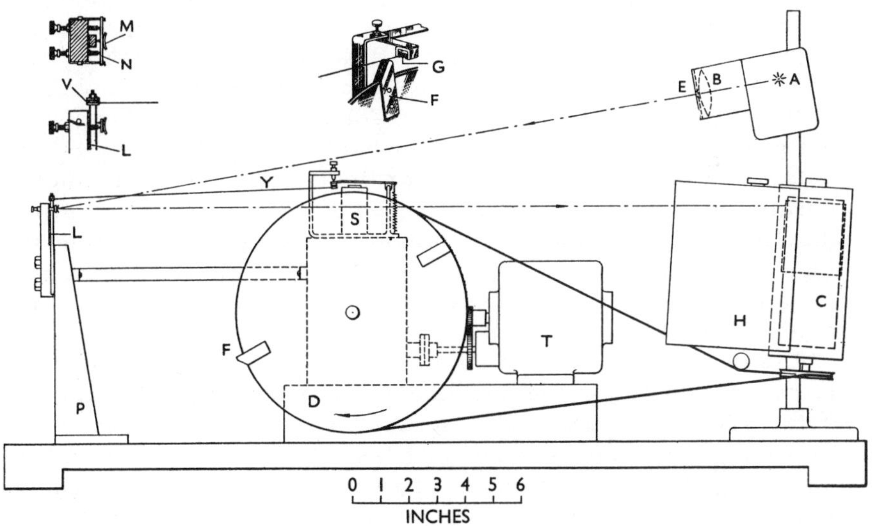

Fig. 3.10. Rotating-disc impact tester used by Meredith.[63]

its width is tapered to reduce inertia.   The yarn (Y) is held between a brass washer and a vulcanized fiber washer (V) to minimize grip breaks.   A small Duralumin plate (N) carried three
needle points, arranged in the form of an isosceles triangle, with
one needle point bearing on the polished surface of the cantilever,
and the other needle points resting in a conical hole and V-groove
made in the ends of two hardened steel adjusting screws (see in-
set).   The plate carries a concave galvonometer mirror (M) and
is held in place by a rubber band.   The mass of the moving parts
is kept to a minimum to reduce inertia: it is estimated that the
maximum inertia error due to the galvanometer mirror and its
support would be about 0.2% for a 500-gm. breaking load, and the
maximum inertia error due to the mass of the cantilever spring
and grip would be about 0.4%.

"The optical system comprises a source of light, lens, slit, galvanometer mirror, and drum camera.   Light from an 18-watt lamp (A) is focused by a lens (B) on to the concave galvanometer mirror (M), which focuses an image of the horizontal source slit (E) on to fast photographic paper (Kodak RP30) in the drum camera (C).   Louvers (H) prevent extraneous light falling on the sensitive paper during the brief exposure time required to complete a test.   A light-tight screen was fitted over the apparatus when the general level of room illumination was high.

"The rotating disc carries two lugs (F) that strike the light-weight yarn grip (G), and carry it away at the circumferential speed of the disc, when the grip-holder is pulled down into the path of these lugs by a solenoid (S).   At the lower speeds of rotation, the grip-holder may be held down permanently, but at higher speeds it is necessary to allow the disc to reach a constant speed of rotation before lowering the grip holder.   Also, at the highest speeds, closing the solenoid switch by hand would not guarantee that the grip-holder had moved fully into position before the lugs reached it.   Accordingly, a thyratron is incorporated in the circuit, controlled by a switch on the rotating disc in such a way that the solenoid is energized just after the lugs reach top dead centre.   This allows the duration of almost a complete revolution of the disc for the armature above the solenoid to close. . . .

"As described above, a beam of light moves in a vertical plane to record tension in the specimen; the drum of the camera is rotated by a cord passing round the rotating disc to record extension at a magnification of just less than unity, so that a load extension curve is recorded.   The camera is tilted so that the relation between the displacement of the light beam on the recording paper and the displacement of the cantilever is linear over the 6-cm. length of drum used in these tests.

"The magnification produced by the optical lever is about 1140 $\times$, so that the grip on the end of the cantilever moves less than 0.15 mm.* for a full-scale deflection of 6 cm., that is, less than 0.05% of a 13-cm. test length.   Accordingly, by traversing the other grip at a constant speed, a constant rate of [elongation, and hence] extension is obtained."   The rate of elongation, the impact velocity, is given by the expression

$$v_0 = 2\pi r R \tag{3.18}$$

where r is the radius of the path of the impact point on the disc, and R is the rotary velocity in revolutions per unit time.

---

* This value should evidently be 0.06 mm, to agree with the magnification and other values given in this sentence.

The photographic record that Meredith obtained was a force-time curve, for, by static calibration of the cantilever, the deflection of the light beam could be readily expressed as force. Since the rate of extension was constant in any one experiment, an extension scale could be fitted to the time axis.   From the known linear density of the specimens, it was thus possible to plot tenacity-extension curves at various rates of extension, which was the form in which Meredith expressed his results.

While, in principle, rotating-disc devices are capable of over-reaching gravity instruments in the impact velocities they can provide, the highest velocity used by Meredith, in his reported work, was only about 140 cm/sec (4.6 ft/sec).   This velocity could be easily obtained with a pendulum or falling weight, since it would require a height of drop of only 4 in.

3.4.3 Method of Stone, Schiefer, and Fox.   Set up at the National Bureau of Standards is rotating-disc equipment with which can be obtained longitudinal impact velocities far exceeding any that could be obtained with a practical gravity device.[104] The central features of this apparatus are shown schematically in Figure 3.11.   The specimen (yarn, cord, or heavy monofil) is

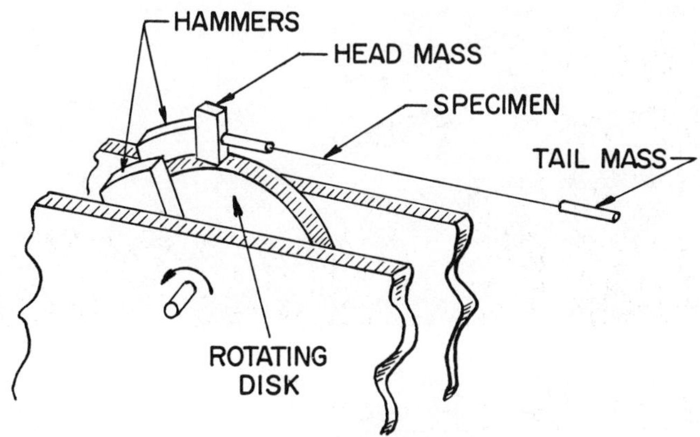

Fig. 3.11. Simplified representation of detail of rotating-disc apparatus designed by Stone, Schiefer, and Fox.

attached at its ends to cylindrical pieces of metal, as shown in Figure 3.12.   One of these, called the "head mass," has a crossbar on it.   This assembly is mounted in a transparent tube, and placed in a horizontal position so that the head mass lies just above the top of a massive disc, which is in the same vertical plane as the specimen.   The disc is rotatable and carries on its flat surface two hammers extending beyond the periphery of the disc. At the outset of a test, the crossbar on the head is in a vertical position, so that the hammers of the rotating disc pass on either side of it, as indicated in Figure 3.11, and the upper view of Figure 3.13.

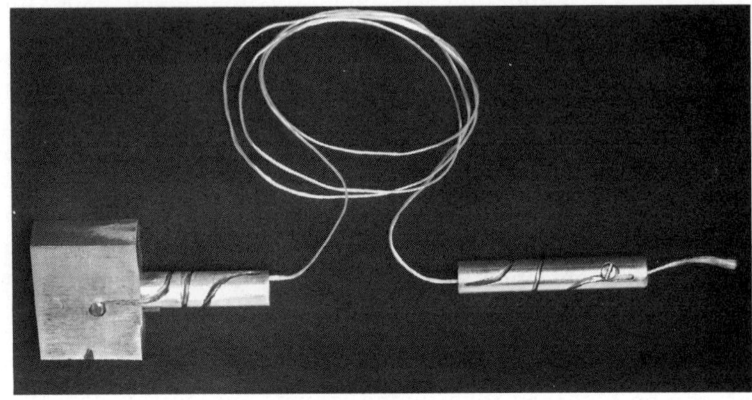

Fig. 3.12. Head and tail masses, attached to yarn specimen, used in Stone, Schiefer, and Fox apparatus.[104]

Fig. 3.13. Head mass and hammers on rotating disc, with head mass in initial position (top), and in position just before impact (bottom).[104]

When a test is to be made, the disc is set in rotation by a variable-speed motor.   When the desired velocity has been reached, the transparent-tube-and-specimen assembly are very rapidly turned through 90°, axially, by electromechanical means, through an intricate triggering circuit.   The crossbar of the head mass is thus brought into position to intercept the hammers of the rotating disc, as shown in the lower view of Figure 3.13. The head is struck on the subsequent passage of the hammers, and is thereby driven forward.   The resultant motions of the head and tail (the mass at the other end of the specimen) are photographically recorded by a high-speed camera.

A photograph of the assembled equipment is shown in Figure 3.14.   In the foreground is the high-speed camera, which is connected to the triggering circuit. When the film attains the proper

Fig. 3.14. Rotating-disc impact apparatus designed by Stone, Schiefer, and Fox; over-all view, showing impact unit, camera, and mirror system.[104]

speed, an electric switch is closed, thereby rotating the head mass into impact position. Images of the head and tail masses, and a centimeter scale, before and after impact, are focussed on the film through a system of mirrors, as indicated in Figure 3.14.   The images appear one above the other in each frame, as shown in Figure 3.15.

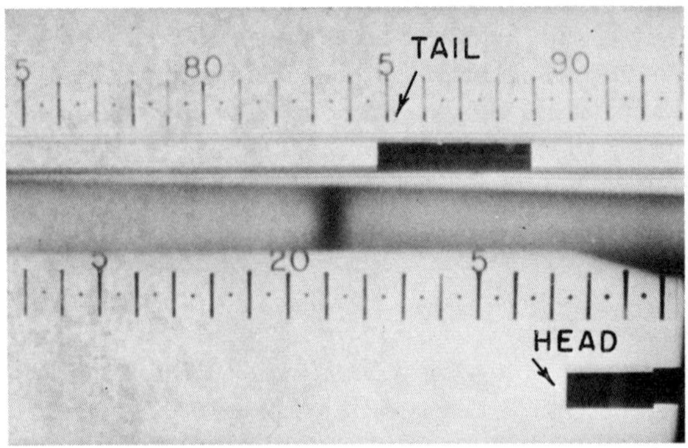

Fig. 3.15. View of field on photographic film, showing head and tail masses before impact.[104]

The sequence of pictures obtained in a typical experiment is shown in Figure 3.16. These were taken at 2630 frames/sec. Impact occurred between the second and third frame; the events recorded in the subsequent 17 frames shown, evidently occurred in about 6.5 millisec. The elastic impact of the metal hammers on the metal head causes the latter to bounce away from the hammers. This can be seen in the lower halves of frames 6 and 7 (where the hammers can be seen coming into the field of view), and in the following three or four frames.

From the positions of chosen points on the head and tail with reference to the scale, as shown by these film records, the distance travelled by these masses could be plotted as a function of frame number, and hence, of time. Typical such curves obtained by Stone and associates are shown in Figure 3.17. Writing of their interpretation of this particular graph, and the velocity-time curves which can be derived therefrom, these authors say:

"The distance-time curve of the head [A of Figure 3.17] is slightly concave downward, indicating that the head is slowing down. The slope of this curve at any point (time) is equal to the velocity of the head at that time. The velocity-time curve of the head is plotted in [Figure 3.18]. Immediately after impact the velocity of the head is about 45 m./sec. At 4 millisec. after impact, it is reduced to 30 m./sec.

"The distance-time curve of the tail [B of Figure 3.17], is concave upward, indicating that its velocity increases with time. At 3 millisec. the slope of the distance-time curve for the tail is equal to that for the head, indicating that the velocities of the head and tail are equal. The velocity-time curve of the tail is also plotted in [Figure 3.18]. The velocity of the tail is zero before impact, then increases slowly immediately after impact. It increases most rapidly at 3 millisec. after impact when it attains

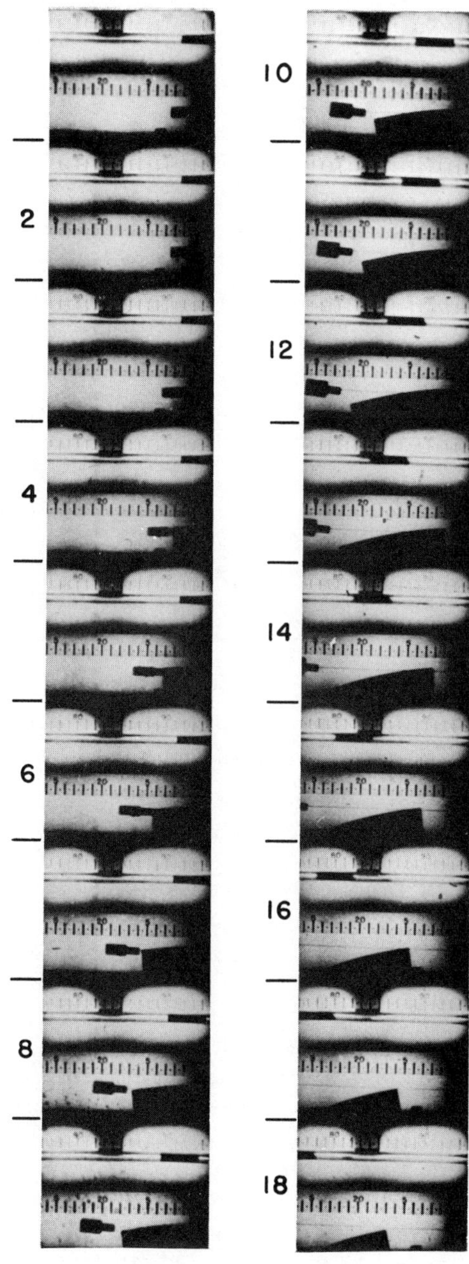

Fig. 3.16.   Film strip showing motion of head and tail masses following impact, which occurred between frames 2 and 3, with maximum tension between frames 10 and 11.[104]

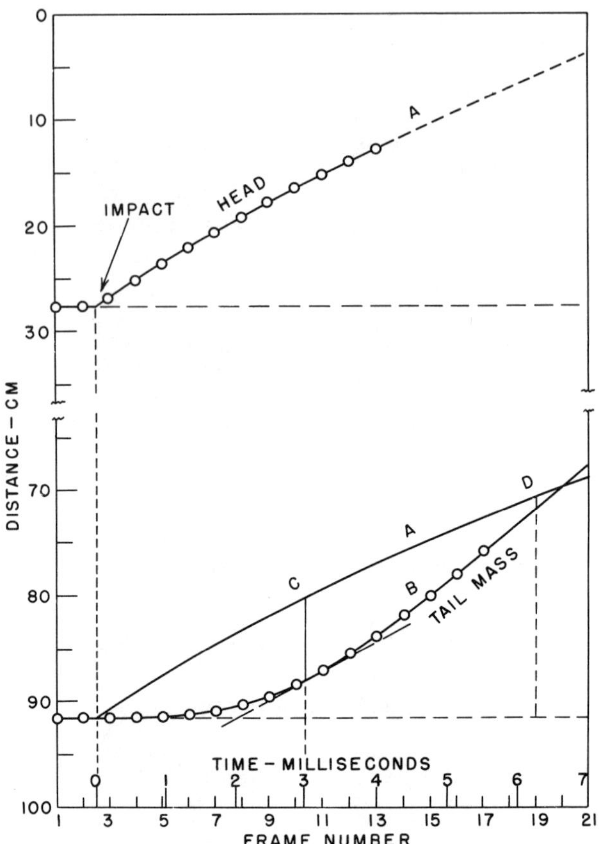

Fig. 3.17. Graphs of positions of head and tail on successive frames following impact of a nylon yarn.[104]

the head velocity of 35 m./sec. It then increases less rapidly,
becoming essentially a constant, 54 m./sec. at 5 millisec." When
the head and tail velocities are equal, at 3 millisec., the elongation and tension in the specimen are at their maxima.   At
later times, the tail starts to overtake the head, so that the specimen tends toward becoming slack.

As in the case of the velocity-time curves similarly derived
from film records by Lyons and Prettyman (discussed in Section
3.2), the slopes of these curves at any instant give the accelerations of the bodies to which they apply.   Stone and associates
computed, in this manner, the acceleration of the tail, in particular. The resulting curve is shown in Figure 3.19.   The
force acting on the tail at every instant (to give it the indicated
acceleration) is given, according to Newton's second law, by the
relation:

$$F = nma_t \qquad\qquad (3.19)$$

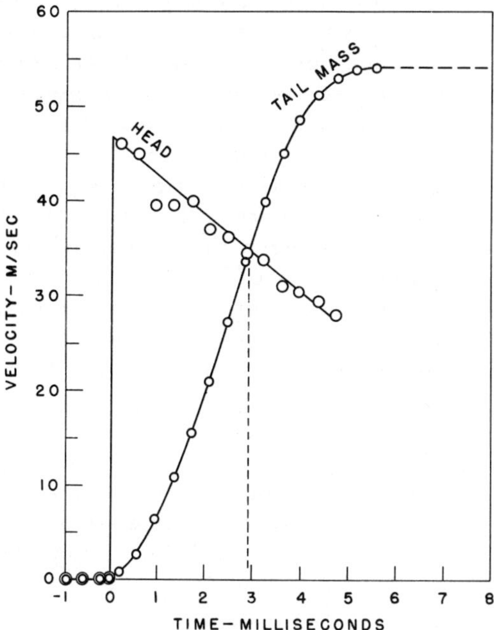

Fig. 3.18. Velocities of head and tail, as functions of time after impact of nylon yarn.[104]

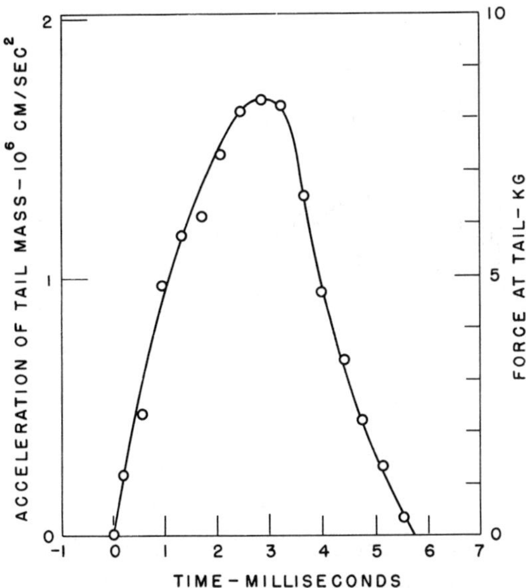

Fig. 3.19. Acceleration of, and force at tail, as functions of time after impact of nylon yarn.[104]

where nm is the mass of the tail, and a_t is its acceleration. This is the force, then, with which the rear end of the specimen acts on the tail, except for the negligibly small additional force required to overcome air resistance. By multiplying the accelerations by the constant nm, one can thus devise a scale so that the acceleration curve also represents the force or stress at the tail end of the specimen, as a function of time. Such a scale has been applied to the right-hand ordinate in Figure 3.19.

The vertical distances between the curves A and B, in Figure 3.17, are evidently the lengths of the specimen at various times after impact. The differences between these distances, and that at impact represent the instantaneous elongations. These are given graphically by the distances between the curves A' and B, where A' is curve A transposed downward so that A' and B coincide for the period up to the instant of impact. From these data, expressing the changes in length as percentages of the original length of the specimen, Stone and colleagues have been able to plot curves of average or over-all extension as a function of time, such as that shown in Figure 3.20.* From the force-time

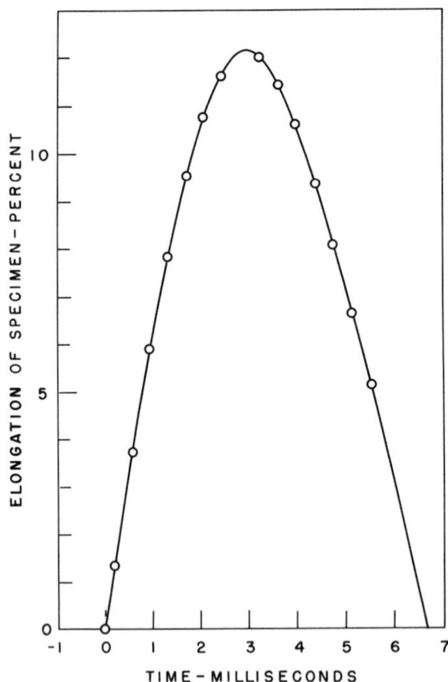

Fig. 3.20. Average extension (elongation) of specimen of nylon yarn, as a function of time after impact.[104]

---

*Stone and colleagues use the term "elongation" in the relative sense in which "extension" is used in the present work.

and extension-time curves, force-extension relationships can
be developed, as has been done essentially by a number of the
other workers mentioned above. The fuller analysis of the results that have been obtained on this National Bureau of Standards
apparatus, and the interpretation of the force-extension (or stress-
strain) curve obtained is presented in Chapter 4.

With this equipment, effective impact velocities in the range
of 10 to 100 m/sec are attainable. Corresponding rates of extension may range up to 5×10^5 %/min.[87,104]

<u>3.4.4 Method of Parker and Kemic</u>. Apparatus that has been
devised for rupturing cords in transverse impact[31,74,83] is shown
schematically in Figure 3.21. Unlike the rotating-disc methods
described above, this one involves the change in angular velocity
of the disc on rupturing a specimen. The specimen, held in rigid

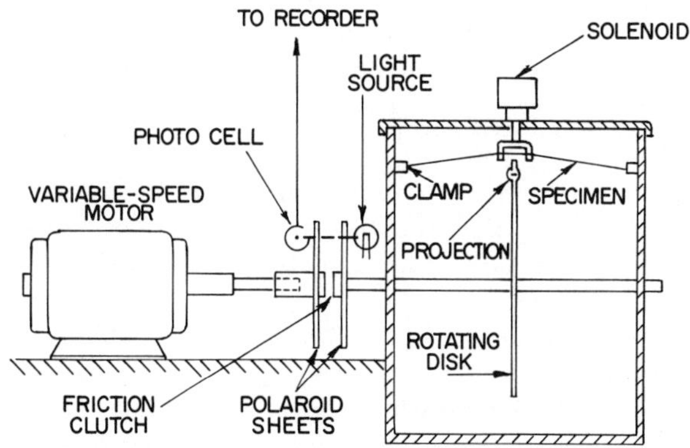

Fig. 3.21. Rotating-disc apparatus of Parker and Kemic for
transverse impact of yarn. [After Reference 83.]

clamps at either end, is supported at its midsection by a two-
pronged fork or hook, above the rotatable disc and perpendicular
to its plane. In turn, the fork, which is vertically movable, is
held in its raised position by an electric solenoid. The disc is
driven by a variable-speed motor through a friction clutch. At
one point on the periphery of the rotating disc is a projection,
which clears the specimen in its raised position, as shown in
Figure 3.21. In making a test the disc is brought up to the desired constant speed, the clutch is disengaged, and the solenoid
is made to release the fork, so that the specimen falls into the
path of the projection on the rotating disc, and is broken.

The quantity measured by this technique is the energy absorbed
in breaking the specimen. This is obtained directly by measurement of the change in speed of the disc before and after impact.
For this purpose two Polaroid sheets are used, one mounted on
the motor axle near the clutch, and the other one mounted on the

disc axle on the other side of the clutch. A beam of light from
a suitable source passes through the two Polaroid sheets and
falls on a photoelectric cell. So long as there is no relative motion between the two sheets, as when the clutch is engaged and
the sheets are rotating at the same velocity, the intensity of light
reaching the photoelectric cell will be constant, and the electric
signal from the cell to a recorder will be constant. When, however, the clutch is disengaged and there is a difference in velocity between the two Polaroids (due to friction in the rotating disc,
or to energy loss in breaking the specimen) the light intensity at
the photoelectric cell will fluctuate cyclically, with a frequency
equal to the difference in circular velocity between the two axles.
The signal to the recorder will vary with the same frequency, so
that the difference in the velocities of the two axles can be measured on the chart. From this difference and the known constant
speed of the motor, the velocity of the disengaged rotating disc
can be readily computed. The energy A_b lost in rupturing a
specimen can then be calculated from the relationship:

$$A_b = \tfrac{1}{2} I_d (\omega_0^2 - \omega_1^2) \tag{3.20}$$

where I_d is the moment of inertia of the rotating-disc-and-axle
system, ω_0 is the angular velocity of this system just before impact, in radians per sec, and ω_1 the angular velocity just after
impact.

In what seems to have been a later modification of the apparatus, the right-hand cord clamp in Figure 3.21 was mounted on a
cantilever beam.[74] The deflection of this beam is sensed by a
linear variable differential transformer connected to the vertical-
deflection plates of an oscilloscope. Thus, it is possible to obtain a photographic record of the load applied to the specimen
during an impact test.

Parker and Kemic have reported that disc velocities corresponding to impact velocities from 10 to 80 ft/sec have been
employed.

3.5 Ballistic Methods

3.5.1 Background. To determine the performance to be expected of textile materials under impact at high velocity, by free-
flying, ballistic projectiles, or missiles, experimental methods
that are themselves ballistic have been developed.   In most of
these, the projectile is accelerated and put into flight by being
driven through the bore of a gun by an expanding gas. In all the
methods, except one (in which the projectile itself does not strike
the specimen), the textile test-piece is subjected to transverse,
rather than longitudinal impact. Both yarns and fabrics have been
tested and studied by one or the other of the ballistic methods.

<u>3.5.2 Test for the V_{50} Limit</u>. A procedure employing a ballistic projectile has been developed by the U.S. Army Ordinance Corps[111] for the acceptance testing of fabrics for armor vests.   The property which this test is intended to evaluate is the ability of a fabric to resist penetration by a free-flying, fragmentary missile.   The criterion of this property is taken to be the lowest (limiting) missile velocity at which complete penetration of the fabric test panel (emergence of the projectile) will barely occur. Because of sample nonuniformity and variability in the test procedure, no absolutely reproducible value for this limiting velocity can be found.   Hence, a statistical quantity, called the V_{50} <u>ballistic limit,</u> is determined; it is interpretable as the striking velocity at which 50% of the individual impacts will result in complete penetration.

A schematic diagram of the experimental arrangement is shown in Figure 3.22, and a photograph of typical range facilities, in Figure 3.23.   The projectile used is a blunt- (though not flat-) nosed cylinder of hardened steel, having a caliber of 0.22 in. and a weight of 17 grains. It is intended to simulate a piece of shrapnel, or a fragment of a grenade. The projectile is fired from a rigidly mounted rifle by means of an explosive, propellant powder.   In the path of the projectile at a distance of 13.5 to 14 ft in the standard method, a panel of the fabric under test is set up, so that the trajectory of the projectile is normal to the surface of the fabric. Behind the panel, as viewed from the gun, is a sheet of Dural (aluminum alloy).   Both are mounted in a rigid frame.

Between the gun and the test panel, in the line of flight, are two frames, known distances from each other and from the test panel. These frames, called "lumiline screens," form part of a chronograph system. Each has a photoelectric cell at the bottom

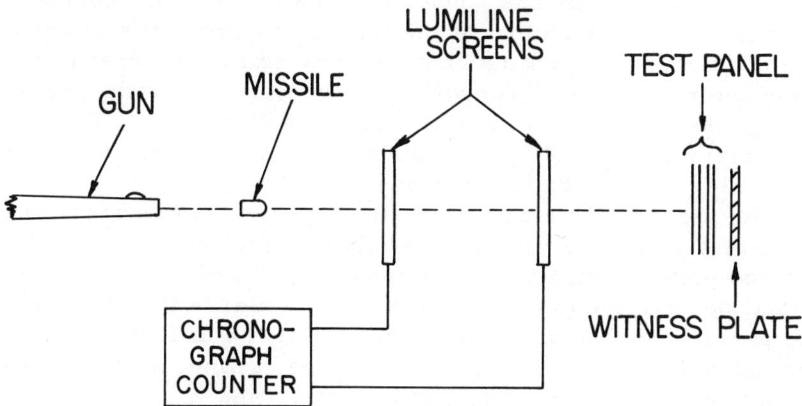

Fig. 3.22. Schematic diagram of arrangement of apparatus for V_{50} ballistic-limit test.

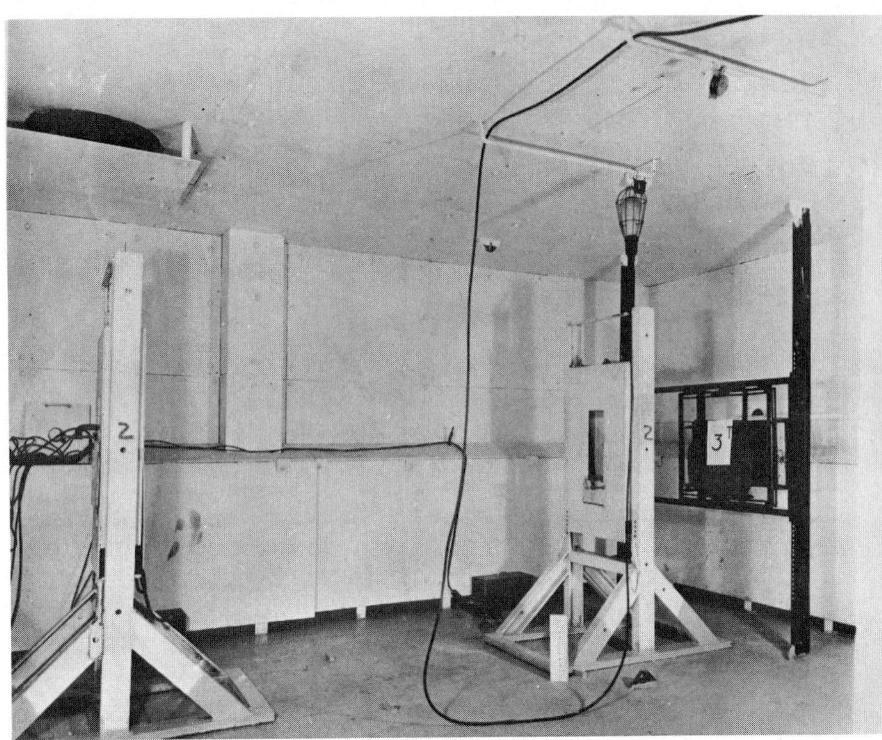

Fig. 3.23. Range facilities for V_{50} ballistic-limit test, showing lumiline screens of chronograph circuit (2), and mounting for specimen panel and witness plate (3).[111]

and a linear source of light at the top, so that there is, effectively, a sheet of light lying across the path of the projectile.
The lumiline screens are connected to an electronic counter,
through a circuit such that at the instant the projectile alters the
beam of light falling on the photoelectric cell of the first screen,
the counter is started; conversely, when the projectile passes
through the second screen, the counter is stopped.   There is
thus obtainable a reading, in millisec, of the transit time of the
projectile between the two screens.

In the practice of this method, the object is to fire projectiles
at the test panel at a variety of striking velocities, so as to obtain complete penetration of the target in 50% of the cases, and
partial penetration (stoppage of the projectile) in the other 50%.
The panel is shifted after each firing so that a different point of
impingement is exposed to the missile each time. A new missile
is used in each firing. The striking velocities v_s are calculated
from the chronograph data for each firing according to the formula

$$v_s = \frac{d_1}{t} (1 - kd_2\rho_{air})$$

(3.21)

where d_1 is the distance between screens, t is the transit time between screens, k is the drag coefficient of the projectile, d_2 is the distance from the midpoint between the screens to the face of the target, and ρ_{air} is the relative density of the air.   The variation of the striking velocity upward and downward, so as to obtain complete penetration in 50% of the cases, is achieved by altering the powder charge, on the basis of experience.   The criterion of a complete penetration is that the projectile pass through the test panel with sufficient residual energy to puncture the Dural witness plate, 6 in. behind the panel. If the projectile is lodged in the panel, or passes through it with insufficient residual energy to puncture the witness plate, the penetration is considered partial.

Firing at the test panel is continued until at least five complete and five partial penetrations (at striking velocities lying within a range of 125 ft/sec) are obtained. Generally, because of the inevitable nonuniformity mentioned above, the results in this range will be mixed; there will be some partial penetrations at higher striking velocities than the lowest for complete penetration. The mean of the five lowest velocities giving complete penetration and of the five highest velocities giving partial penetration is taken as the V_{50} ballistic limit.

In the acceptance testing of the nylon fabric currently (1961) used by the U.S. Army for body armor[112,113] for ballistic resistance, the test panel consists of twelve layers of fabric, about 15 in. by 15 in. For such a panel, when tested with the designated .22-caliber projectile (T-37), a V_{50} limit of not less than 1225 ft/sec is specified.   In research and development studies, of course, other combinations of fabrics may be used.* The areal density of an experimental panel (its total mass divided by the area of the surface exposed to impact by the projectile) is, however, always taken into consideration in assessing the significance of the ballistic limit. An analytical discussion of the relation of the V_{50} limit to areal density is given in Section 4.2.

3.5.3 Method of the Chemical Warfare Laboratories. Ballistic techniques that grew out of those of the test for the V_{50} limit were developed by the Chemical Warfare Laboratories (C.W.L.) of the U.S. Army for research studies on textile materials under high-speed impact.[80,52] An outstanding feature of these studies was the use of high-speed photography to observe the behavior of both fabric assemblies and single yarns during the impact process. The arrangement of the ballistic facilities is essentially the same as those for the V_{50} test, shown in Figure 3.22, though the gun-to-target distance is much shorter. Both a conventional, powder-propellent gun, and one in which the propellent is rapidly expand-

---

* For such studies, fragment-simulating projectiles in a number of weights from 1.5 to 207 grains are available.

ing helium, have been used. The same .22-caliber projectile that is used in the V_{50} test has been used with these guns. Instead of lumiline screens, electrically conductive, metallic grids, printed on thin paper, are used to start and stop a chronograph counter. Passage of the projectile through the grids breaks them; the resulting loss of electrical contact appropriately actuates the counter circuit. It is stated that, with these facilities, velocities could be measured to less than 0.2% error and times of impact to ±2 microsec.

For the high-speed photography, cameras are set up at the target to view it along a plane perpendicular to the line of flight of the projectile (the plane of the photographic film is parallel to to the line of flight). In the first arrangement, which was for the study of the behavior of panels of armor fabric, a motion-picture camera providing about 15,000 frames/sec, was used. A light source illuminated the back of the test panel.   By this means a photographic record, at very short intervals of time (fractional milliseconds), of the progressive deformation of the back of a panel was obtained. In another arrangement a micro-flash technique was used; the back of the test panel, with a camera focussed on it, was illuminated once, for a fraction of a millisecond, at some predetermined time after impact.   This involved use of an electronic interval generator which, triggered by rupture of the second chronograph grid, produced a pulse that actuated the light source, after an accurately known delay. From measurements on photographs of a series of specimen panels of a fabric sample, it was thus possible to plot the displacement of the point of impact as a function of time.[52]

A similar microflash technique was used in these laboratories for the study of the displacement and longitudinal strain of nylon yarns impacted transversely by the same projectile.[79,80] As finally developed, the method employed a comparatively long specimen (120 cm), to avoid the complicating effects of the reflection of strain fronts from the clamped ends. This required, in order to get a photographic record of the middle (impact) and lower portions of the yarn, the use of three cameras, one above the other, in about the same position as in the fabric experiments.   The yarn had narrow black "tick" marks, spaced at equal distances, placed on it, for purposes of these studies. In carrying out an experiment, the yarn, hanging under a small load, was photographed before being impacted. The camera backs were then slightly shifted manually, without removing the film.   The projectile was fired at the yarn, with the same sequence of events as in the fabric case. In the initial experiments a single micro-flash exposure was made during the transverse displacement of the yarn by the projectile.   It was found that the transversely displaced portion of the yarn assumed a very sharp V-shape, with the projectile at the apex. This V expanded along the yarn as the projectile moved forward. The junctions of the V with the undeflected segments of the yarn have been called transverse waves.

A typical set of photographs from the three cameras is shown in Figure 3.24, covering the yarn from a point above the impact region A to the bottom clamp C. Measurements were made on the negatives with a microprojector, which throws onto a ground-glass screen an enlargement of the image on a film. From measurements of the position of the transverse wave relative to the impact point, on films exposed known times after impact, the

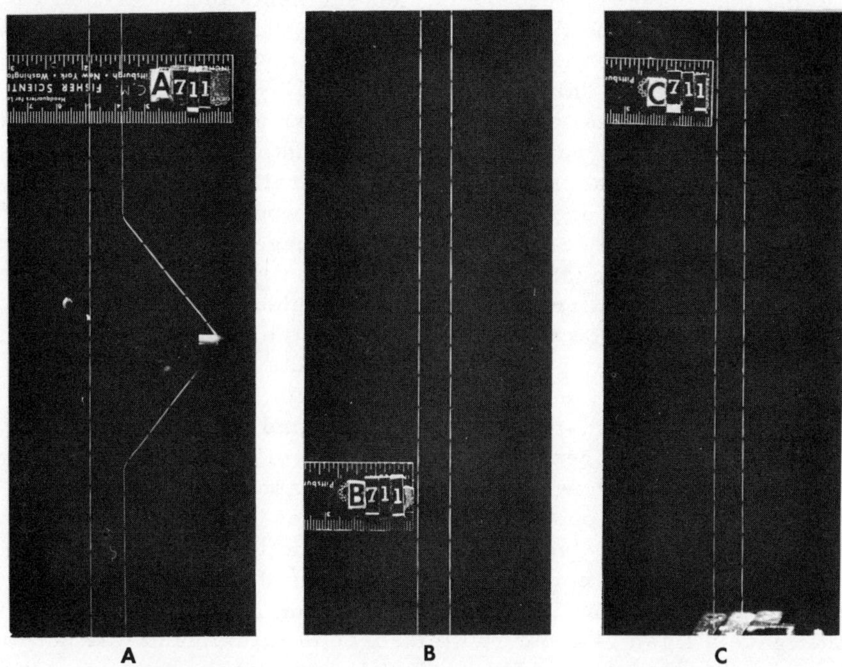

Fig. 3.24. Photographs of yarn specimen shortly after impact, obtained by method of the Chemical Warfare Laboratories.[80]

transverse-wave velocity could be calculated.   Thus, it was possible to establish the relationship between this velocity and striking velocity. By measuring the distances between tick marks along the images of the yarn, before and after impact, the tensile strain in the yarn at various points could be found. These measurements also gave the position of the strain front or wave (the point ahead of which the yarn is yet unstrained)* at the known

* For an explanation of the strain-propagation phenomena occurring here, the reader is referred to Section 4.1.

instant the photograph was made. Since the time after impact was known, the strain-wave velocity could also be calculated.

The microflash method described in the foregoing two paragraphs, required a separate shot, on a different specimen in each case, in order to get a record of the strain in a yarn as a function of time after impact. It evidently would be preferable if the strain history could be obtained on a single specimen with a single shot, at one striking velocity. To accomplish this, Jameson and associates, in a later arrangement,[39,40] made use of three microflash units, each set, through its own interval generator, to flash in sequence at selected intervals. To avoid having the images of the undeflected segments of the yarn at each stage overlap on the film, thus obscuring the tick marks, it was necessary to find some means of slightly displacing the images from each other. This was accomplished by placing a rotating mirror in the optical path between the yarn and the cameras. The cameras, in this arrangement, were mounted so that their lines of sight, directed at the mirror, were parallel to the trajectory of the missile. In order that the mirror would be in the proper positions to reflect to the cameras the images of the yarn at each instant that the microflash units went on, firing of the projectile was timed by the mirror itself. To achieve this a mercury switch, having contacts that rotated with the mirror, was employed. The proper position of the contacts relative to the mirror was computed from the mirror velocity, the expected projectile velocity, gun-to-target distance, and the delay times of the system. A representation of the type of record obtained by this technique is shown in Figure 4.9, Chapter 4.

These ballistic, photographic procedures were further elaborated with the use (sometimes in conjunction with the multiflash technique) of continuous illumination of the yarn during the impact process. The illumination was from a studio speed lamp that gave a flash of about 200 microsec duration, and was set to be triggered just before the projectile struck the yarn. The image of the illuminated yarn was swept across the photographic film by the rotating mirror, thus tracing the loci of the tick marks at every instant during the impact process. From measurements on these loci, the longitudinal strain at various points in the undeflected portions of the yarn, at any instant, could be calculated.

A final development of the C.W.L. technique for yarns is in use at Fabric Research Laboratories.[5,43] The unique feature is the use of an Edgerton high-speed stroboscopic light source that is capable of producing 15 accurately timed, consecutive flashes, each of very short duration. This unit replaces the bank of two or more single-flash units originally used at the C.W.L. Facilities for launching the projectile, the position of the camera, and

so forth, are substantially the same. The equipment provides,
on a single film, a rather detailed record of the position of the
V-shaped deflected section of the test yarn during the rupture
process.

3.5.4 Method of the National Bureau of Standards. Smith and
colleagues[100] assembled apparatus for what may be considered
a ballistic method in that a specimen yarn was struck transversely
by a free-flying projectile, though the means of propulsion are
not conventionally "ballistic" in the modern sense.   These authors describe their equipment (a photograph of which appears in
Figure 3.25) as follows: "The yarn specimen [S] is clamped to
a rigid massive table, A, on which a coordinate grid system is
inscribed. Central transverse impact is made by a freely flying projectile that has been struck by a rapidly rotating hammer,
H. Apparatus for rotating and stopping the hammer is separate
from the specimen table in order to avoid jarring.

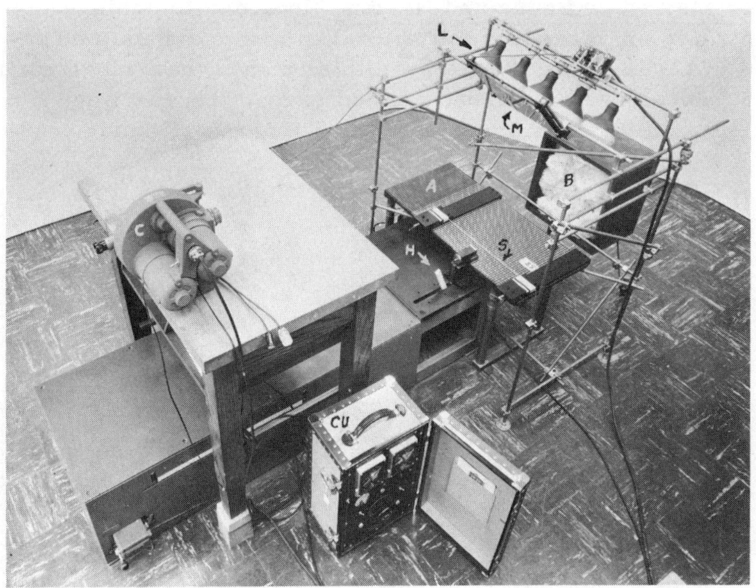

Fig. 3.25. Apparatus of the National Bureau of Standards
for transverse impact of yarn.[100]

"The 6-in. hammer rotates under a powerful torque through
an arc of 270° before striking. The force to rotate the hammer
is applied at the surface of a 2-in. shaft by straps from four
springs, which can be extended up to 20 in. by a motor.   At full
extension the total tension in the four springs is 800 lb. The hammer is held in place by a latch, which can be suddenly released by
a solenoid. Projectile speeds have been measured as high as 70m./sec.

"Reflected images of the specimen after impact are photographed by the high-speed camera, C, on the table. Either 7,000 or 14,000 pictures per second can be taken, depending on which of two Fastex cameras is used for this purpose.

"Other parts of the equipment shown [in Figure 3.25] are the control unit, CU, which puts timing pips on the film and triggers the hammer when the camera is up to speed, the five 750-w flood lamps, L, for illuminating the specimen, the mirror, M, for reflecting the image of the specimen into the camera, and the box, B, for catching the projectile."

A typical photographic record that is obtained, showing the position of the yarn specimen at consecutive intervals of about 0.7 millisec, is shown in Figure 3.26. With the use of a microfilm reader, the positions of the projectile, the transverse-wave front, and the clamps, in each frame, are measured. How Smith and associates use these data to derive the stress-strain curve of a yarn sample is described in Section 4.1.

3.5.5 Method of Lewis and Holden._ In order to achieve higher strain rates on yarns and monofils than had previously been attained in his laboratory by Meredith's method (described earlier), Lewis[48] developed a ballistic method to load the specimen longitudinally. In this technique the free-flying missile did not strike

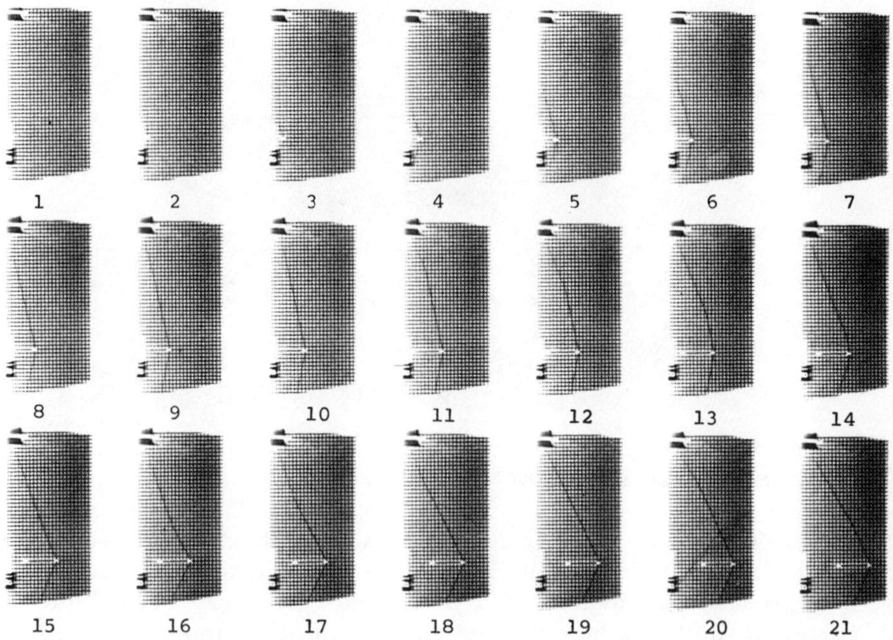

Fig. 3.26. Film strip showing motion of projectile and successive configurations of yarn specimen after transverse impact, which occurred in upper left-hand frame.[87]

the textile specimen, but served as a means of rapidly accelerating a rotatable arm to which the specimen was attached.   A
schematic diagram of the apparatus is shown in Figure 3.27.

  The specimen was mounted on one face of a piezoelectric stress
gage, the other face of which was attached to a fixed horizontal
steel rod. The other end of the specimen was attached to the upper end of a vertical arm, which was rotatable about a fixed pivot
at its lower end. The small (0.95 g) projectile that was used was
fired from an air rifle at an anvil midway along the arm, with a
striking velocity of about 500 ft/sec. The resulting tensile force
applied to the yarn was detected by the stress gage, and through an
electronic circuit, was registered on the Y_1 channel of a double-
beam, cathode-ray oscilloscope. There was thus obtained by photography, on the basis of a prior calibration, a record of the tension in the specimen as a function of time during the impact process.

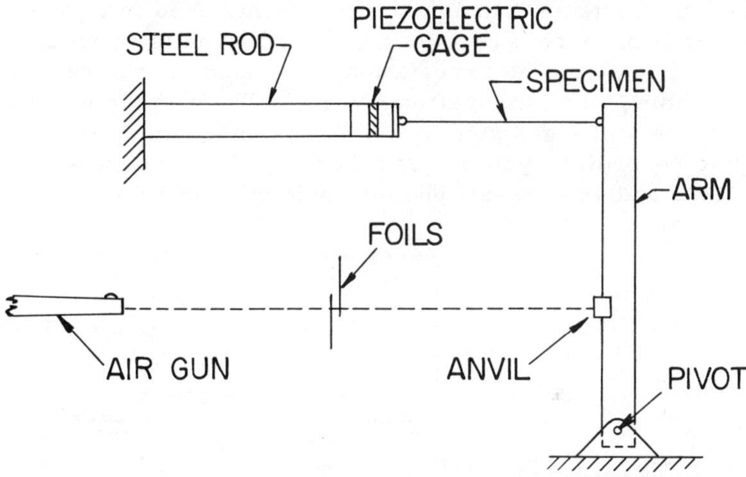

Fig. 3.27. Schematic diagram of ballistic apparatus designed
by Lewis for impact tests on yarns. [After Reference 48.]

  The rate of extension of the specimen, which was considered to
be constant, was given by the interruption, by the swinging arm,
of two beams of light falling on the cathode of a photoelectric cell.
This cell was connected to the Y_2 plates of the oscilloscope, so
that as the arm interrupted first one, and then the other light beam,
discontinuities appeared on the oscilloscope tracing, which could
be photographed. The sweep on the time (X)-axis, for both tracings, was triggered by the momentary contact between two foils
when they were punctured by the missile. From the known distance between the light beams and the sweep speed of the oscilloscope (which gave the time between discontinuities), the rate
of extension could be readily calculated. With a record of both
the tensile force and extension of the specimen as functions of

time, the force-extension curve could be plotted in the usual manner. Such data were obtained on nylon, acetate, and Terylene samples at rates of extension above 20,000 %/sec.

Holden[37] modified this apparatus so that it was capable of providing rates of extension up to 66,000 %/sec. Essentially, his remodeling consisted of pivoting the arm (of light hard aluminum) at its upper, rather than lower end; arranging to impact the arm with the missile at a point very near the specimen attachment; and using only one beam of light, carried to the photoelectric cell through a Perspex, internal-reflection light guide, to measure extension.

<u>3.5.6 Method of Fabric Research Laboratories</u>   Among the most massive ballistic apparatus for the impact testing of textiles is that described by Chu, Coskren, and Morgan.[18]   The equipment is said to have been built to study the impact behavior of such products as "parachute components used in missile, nose cone, and space-ship recovery, . . . safety belts, suspension lines, crash helmets and the arrestation nets used to restrain a jet plane landing on an aircraft carrier."[3] The complete installation consists of a gas gun, two ballistic pendulums, and a photographic recording system. A schematic diagram appears in Figure 3.28, and an over-all photographic view in Figure 3.29.

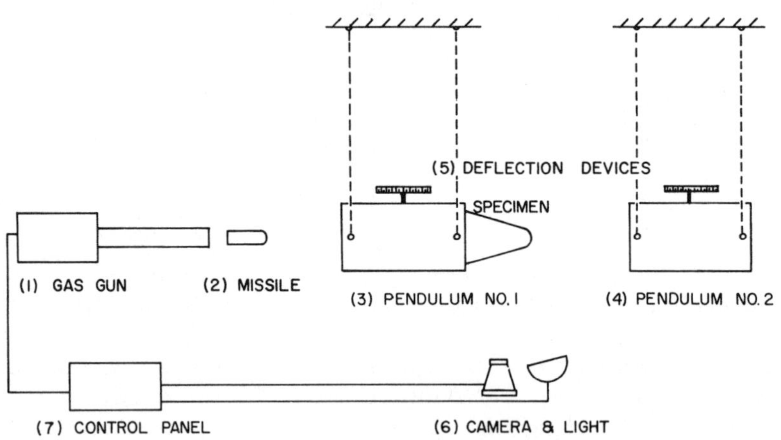

Fig. 3.28. Schematic diagram of ballistic equipment at Fabric Research Laboratories for impact testing of textiles.[18]

In operation, a missile 2.5 in. in diameter is fired from the gun at the specimen, mounted in a V-shape in the path of the missile on the back of the first pendulum. When the projectile strikes the specimen, the pendulum is set in motion. The displacement of the pendulum is recorded on the deflection devices (shown in Figure 3.28), and at the same time, the positions of the missile and of markings on the specimen are photographically recorded at known intervals after impact, by means of an Edgerton stroboscopic light source described earlier in this section.   If, as is usually the case, the specimen is ruptured, the missile passes on to the second pendulum where it is engaged and stopped.   The displacement of this pendulum resulting from the impact is recorded on a scale, as in the case of the first pendulum.

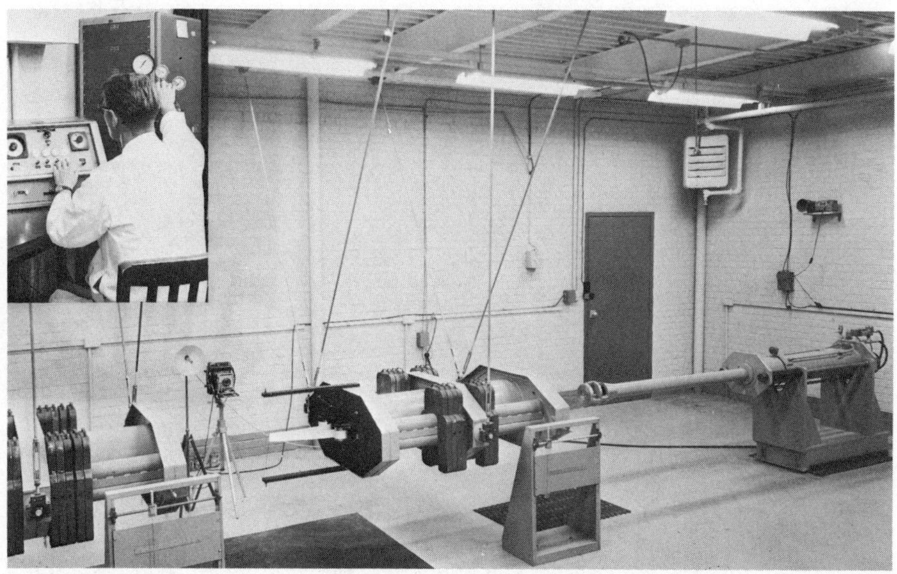

Fig. 3.29. Over-all view of ballistic equipment at F.R.L. for impact testing of textiles; remote-control unit in upper left-hand corner.[18]

In the gun, which has an over-all length of about 10 ft, nitrogen or helium under pressure is used to launch the missile. The latter is a cylinder, with a semicylindrical nose. The nose and tail are of phenolic plastic, while the midsection may be either solid or tubular, and made of plastic or metal, depending on the mass required to deform or rupture the test material. Projectiles of 8, 16, and 40 oz, and 10 lb have been used. Each pendulum consists of a set of cables, hung from the ceiling, supporting an aluminum tube (through or into which the missile passes) fitted with exterior rails for supporting weights that may be added, in order to increase the mass of the unit. Masses up to 1 ton are attainable.

The equipment is operated, by remote control, in a darkened range, with an open-shuttered camera focussed on the test specimen and the impact region. On the basis of calibration curves, the proper gas pressure is preselected to give to the missile, of the particular size being used, the desired striking velocity. The circuit of the stroboscopic, or multiflash light source is triggered so that the flashing starts an instant before the beginning of the impact process.   There is thus obtained, on a single-photographic film, images of a gage mark on the missile, after each interval of time marked off by the stroboscope.   On the same film are also the corresponding images of two or more gage marks placed on the test piece.   The consecutive distances between pairs of images provide a record of the extension of the specimen during the impact process.   A typical photograph of an impact on a specimen of heavy tape is shown in Figure 3.30.

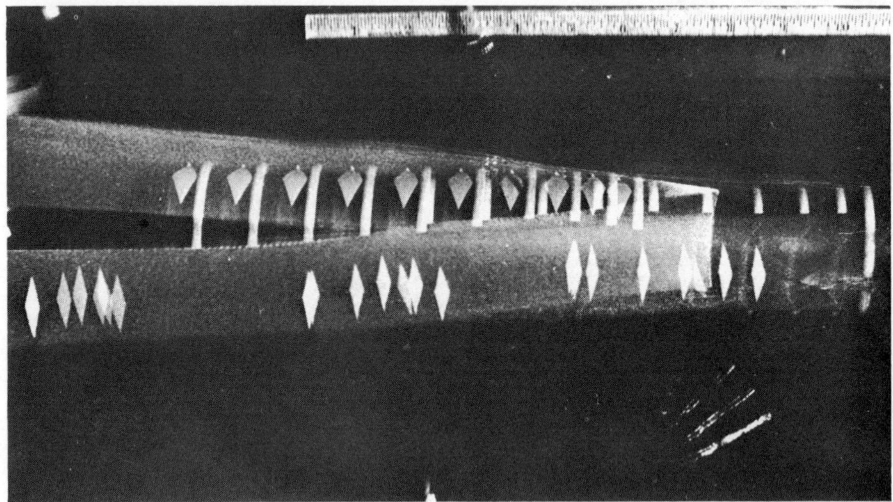

Fig. 3.30. Photographic record of impact process obtained on F.R.L. equipment by multiflash technique.[18]

From measurements on the film record of the successive displacements of the missile (in much the same manner as was done by Lyons and Prettyman[51] for the motion of a pendulum bob, as described in Section 3.2) Chu, Coskren, and Morgan are able to compute the velocity, deceleration, and force of the missile during the impact process, up to rupture. In addition (as indicated in earlier discussion), from measurements directly on the image of the specimen, extension as a function of time can be obtained.   From the force-time and extension-time relationships, a force-extension curve can be readily derived.   Finally, the work to rupture can be obtained either by measuring the area under the force-extension curve, or by calculation from the velocities of the projectile before impact and at, or after, rupture, according to the equation:

$$A_b = \tfrac{1}{2}m(v_s^2 - v_r^2) \tag{3.22}$$

where A_b is the work to rupture, m is the mass of the missile, v_s is the striking velocity of the missile, and v_r its velocity at rupture. These velocities, as indicated in previous explanation, are obtained from the film. The rupture force and work to rupture obtained in this manner may be checked by calculating these properties from the displacements of the two pendulums.

Striking velocities up to 800 ft/sec and rupture forces up to 200,000 lb are attainable. The advantages of this equipment according to Morgan,[3] are: "1) it can test large samples having up to 10,000 lbs. static strength, including not only basic materials, but fabricated structures as well; and 2) precise measurements are possible enabling the accurate calculation of stress-strain data. . . . The latter was not the case in the multi-mile rocket-sled track previously used for heavy impacting of parachute components, and the sled track had the further disadvantage of extra expense and time consumption per test."

3.6 Other Recently Developed Methods

3.6.1 Pneumatic and Hydraulic Apparatus. In recent years have appeared descriptions of equipment employing pneumatic or hydraulic means, or combinations of these, for the impact loading of textiles. The impact velocities obtained are not in the order of those provided by ballistic methods, or even some of the rotating discs, such as that at N.B.S., but they are adequate for many research purposes. The pneumatic and hydraulic devices in general, but especially those having high load capacity, can be more conveniently and safely operated than can falling-weight machines providing the same impact velocity and capacity.

An impact tester, called the MITEX, employing a pneumatic-hydraulic system, has been developed at M.I.T. for the testing of yarns and other textiles at velocities ranging from 1 in./sec to 20 ft/sec. The operation of this tester (as described by Krizik, Mellen, and Backer[42]) depends upon the initial application of the pneumatic pressure of a nitrogen supply to one side of a piston and the hydraulic pressure of an oil to the other side. On the piston rod, the axis of which is vertical, is mounted the lower clamp for the specimen. In the performance of a test, the pneumatically pressurized upper end of the cylinder is connected to the main nitrogen reservoir. The hydraulic fluid in the lower end of the cylinder is then suddenly vented, through a large hand-operated valve, into a reservoir at atmospheric pressure. The resulting rapid downward movement of the piston imposes the impact load on the specimen. A schematic diagram of the pneumatic-hydraulic system is shown in Figure 3.31. The tester has a load capacity of 500 lb.

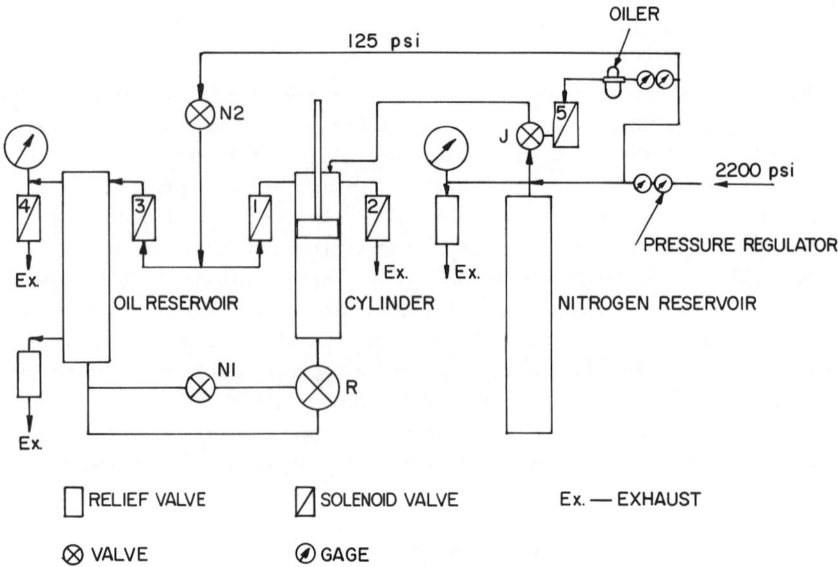

Fig. 3.31. Schematic diagram of pneumatic-hydraulic system of MITEX impact tester. [After Reference 42.]

The upper clamp of the instrument is suspended from a screw jack mounted on the frame of the machine.   By means of the screw jack, the gage length of the specimen may be varied. The methods of measuring and recording the force on the specimen, and the elongation, as functions of time, by means of a piezoelectric transducer, magnetic tape, and an oscilloscope, are very much the same as were used on the M.I.T. falling-weight tester (described in Section 3.3). In the present apparatus the magnetic tape may be attached to either the lower jaw or the specimen.

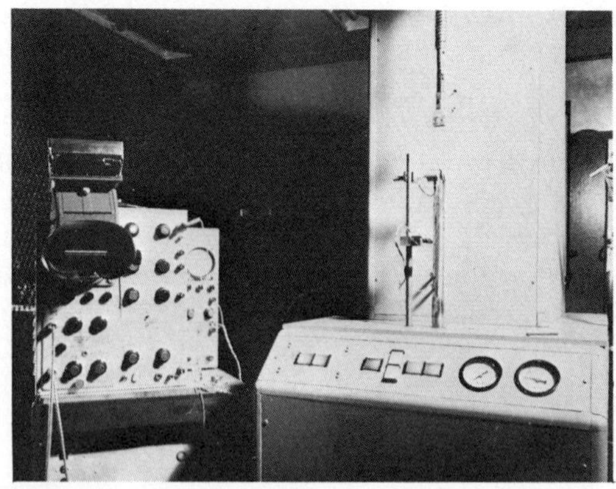

Fig. 3.32. MITEX impact tester, with oscilloscope, camera, and other auxiliary equipment [private communication, Stanley Backer].

A somewhat similar, commercially available tester, called
the Plastechon,[*][105] also has been used in the impact testing of
textiles.[43]   In this machine the hydraulic system is said to be
operable by air, also. Constant loading rates up to 10,000 in./min
are claimed, with a capacity in tension of 10,000 lb, in one model.
Force is detected by means of a strain gage, while elongation is
measured with a potentiometric transducer.   Both signals are
transmitted to the screen of an oscilloscope so as to display
the force-elongation curve of the specimen during the testing
process.   Provisions are made for photographically recording
the curve.

Equipment actuated by a pneumatic system has been built for
the laboratories of the Celanese Corporation of America, for impact testing of single fibers.[†]   A schematic diagram of the apparatus is shown in Figure 3.33.   As in the MITEX tester, the
loading mechanism is a piston, to which through the piston rod,

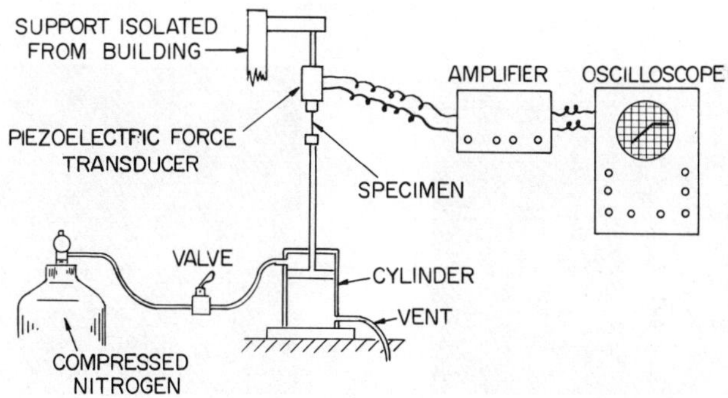

Fig. 3.33. Schematic diagram of pneumatic impact tester
for fibers, at Celanese Corporation laboratories.

the lower end of the fiber specimen is attached.   To minimize
mass and thus achieve higher accelerations, the piston and rod
are constructed of aluminum. The upper end of the specimen is
attached to a piezoelectric force transducer. Because of the extreme sensitivity required of the transducer for measurements
on single fibers, it is sensitive to acoustical noise and vibration.
The transducer is therefore mounted on a support resting on an independent foundation, isolated from the laboratory building.   The
transducer is connected to an oscilloscope, through an amplifier.

In the performance of a test, compressed nitrogen is admitted
to the top of the cylinder, thus driving the piston downward. The
tension induced in the specimen acts on the transducer with a

---

[*] Manufactured by Plas-Tech Equipment Corp., Natick, Massachusetts.

[†] Private communications. [Also Reference 70.]

resultant signal that is proportional to the tensile force. There
is thus displayed on the oscilloscope screen a tracing of this
force as a function of time. Typical photographs of tracings on
the oscilloscope screen that are obtained with this system (and
similar ones) are shown in Figure 3.34.

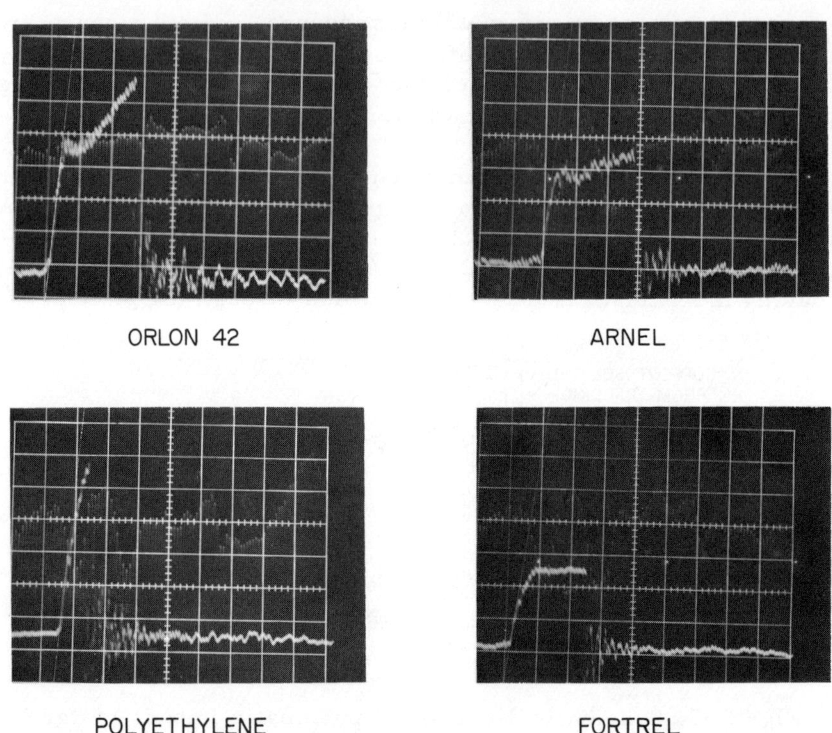

Fig. 3.34. Typical photographic records of tracings on oscilloscope screen for various fibers, at an impact velocity
of 15 ft/sec, as obtained at the Celanese Corporation labo-
atories [private communication].

Calibration of the hydraulic system with respect to piston
speed is achieved by employing prerecorded magnetic tape in
the same fashion as has been used to measure specimen extension on the M.I.T. testers. Data from which force- (or tenacity)-
extension curves can be derived are thus made available.   With
the use of short (1-in.) specimens, and attainable piston speeds
of 30 ft/sec, rates of extension in excess of 10^6 %/min are achieved.
In September 1962, a large impact machine of an entirely different load range, but also powered by a pneumatic system, was
installed in the U.S. Army Laboratories at Natick, Massachusetts,
for the high-speed testing of heavy textile structures such as parachute suspension lines, fabric, and webbing.   Nitrogen is used

in the pneumatic system. An ingenious feature is the use of two cylinders to double the effective rate of elongation at which the impact load is applied. This tensile load is applied horizontally. The specimen is held between the cylinders, in clamps on the two piston rods. Each of the oppositely acting cylinders can apply a load at rates up to 30 ft/sec, so that the total rate of elongation of a specimen can be as high as 60 ft/sec. The minimum rate of elongation is reported to be 30 ft/sec.

Specimens from 10 in. to 6 ft can be accommodated. The full strokes of the pistons allow 35% extension in the longest specimens. The machine was designed to have a load capacity of 6000 lb, and provide a maximum deformation energy of 9000 ft-lb. As in several of the devices described in preceding sections, auxiliary equipment is provided, so that tracings representing force-elongation relationships can be displayed on an oscilloscope screen and photographically recorded.[*]

3.6.2 Rocket Sleds. What may be regarded as the most massive and elaborate arrangements ever to be employed in the impact testing of textiles are the rocket-sled systems set up at the Royal Aircraft Establishment, at Farnborough, Great Britain[14] and at Edwards Air Force Base, in California.[118] These systems were designed originally to simulate on the ground the conditions of high speed and high acceleration encountered in modern aeronautics. Their capabilities have been adapted to a variety of studies, including aeromedical research, ejection-seat development, and parachute engineering. The use of the rocket sled (or rocket-propelled trolley, as the Farnborough equipment is called) for the testing of textile cables or webbing is only an incidental application. Since these installations were designed to do much more than the impact testing of particular types of textiles, they are uneconomical in this application, and hardly qualify as textile-testing apparatus.   Only the equipment and technique used at Edwards AFB will be described here, and that briefly.

The outdoor installation consists of thousands of feet of two-rail track, on which slides a more or less streamlined vehicle, driven by rocket engines. For the tests on the nylon webbing, the sled was equipped with a "bumper," located at the front in such a position that it would engage the specimen, mounted between the rails in a V-configuration, as shown in Figure 3.35.   Between the clamps holding each end of the webbing specimen, and the fixed supports, were strain-gage units for measuring the force on the webbing during the impact process.   The signal from the force-measuring unit was transmitted through electronic circuits to an oscillograph in a remote control room.   In the

---

[*] Private communication, W. E. C. Yelland, Quartermaster R. and E. Command, Natick, Massachusetts, 1 Feb. 1963.

Fig. 3.35. Front view of rocket sled in position of impacting specimen of nylon webbing.[118]

original arrangement, the elongation of the specimen as a function of time seems to have been determined by recording, by
high-speed photography, the position of a pointer on the sled,
with reference to a track-side scale; much the same manner was
employed by others to follow the motion of a pendulum bob during an impact process, as described earlier in this chapter.
Other refinements, such as radio telemetry of force measurements, from the speeding sled, were introduced or projected.

In the reported procedure,[118] for each experiment four specimens were mounted at positions several yards apart along the
path of the sled. The sled was started some distance back on
the track, and was accelerating as it encountered the specimens
at the consecutive locations. In most experiments, the impact
velocity was about 750 ft/sec (500 mi/hr) when the fourth specimen was reached.

3.6.3 Sling-Shot Machine of Pilsworth. An apparatus that bears
some resemblance, in miniature, to the rocket sled, though otherwise powered, has been built and used at the Quartermaster
Research and Engineering Center.[82] A top-view drawing of the
machine is shown in Figure 3.36. A slider, consisting essentially of a steel plate with guiding flanges, is mounted on two
rails about 10 ft long. It is pulled into its starting position by
cables on a winch, against the elastic, retractive forces of two
heavy rubber rings attached to outboard hooks on either side of
the slider. The specimen, about $27\frac{1}{2}$ in. long (in the case of a
nylon yarn that was tested), is mounted between the rails, ahead

of the slider and beneath its path. As in the method of Stone and associates,[104] at the forward end of the specimen is affixed a T-shaped head mass.   While Pilsworth has varied his experimental procedures, in one of these he used a tail mass, in further similarity to the technique of Stone and associates.

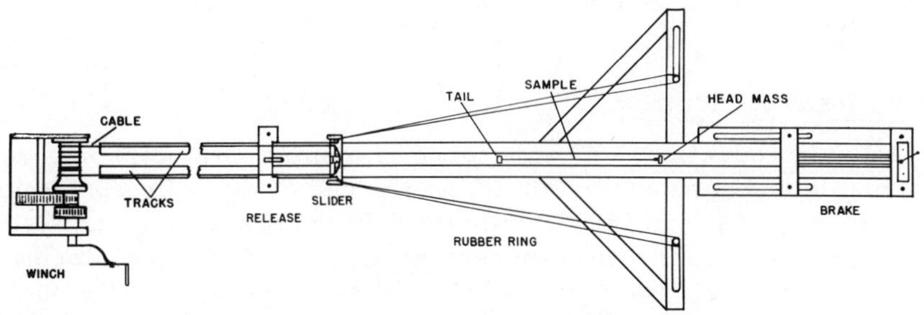

Fig. 3.36. View, from above, of sling-shot machine of Pilsworth.[82]

To conduct a test, the slider is released, permitting it to move forward very rapidly under the retractive force of the rubber rings. Two projections on the bottom of the slider straddle the specimen but engage the projecting crossbar of the head mass, sending this, and the specimen, into flight ahead of the slider, because of the "bounce" effect. Impact velocities in the range of 130 to 200 ft/sec have been employed with this equipment.

Much after the manner of the Chemical Warfare Laboratories technique,[52,80] recording of the strain history of the specimen immediately subsequent to impact is accomplished by means of a system of mirror, camera, and stroboscopic light-source, mounted on a frame above the track and specimen.   For the analysis of the distribution of local strain, tick marks have been used on the specimens.   The treatment of the data when a tail mass is used is similar to that of the N.B.S., as described in connection with Figures 3.18 to 3.20.

Chapter 4

THEORIES OF IMPACT PROCESSES

4.1 Linear Structures

4.1.1 <u>Introduction</u>. When a force is applied at a point in a deformable body, as for example, at the end of a textile specimen in a tensile test, an infinitesimal strain occurs instantaneously at that point. In the next increment of time, through molecular interaction, the strain spreads to adjacent points, while the strain at the point of application tends to increase. Generally, the strain spreads or propagates very rapidly, compared with the rate at which the specimen is elongated* in, for instance, the conventional tensile test, or even some of the so-called impact tests which operate in the lower speed ranges. Thus, it can be said that, on the time scales of such tests, the strain, through direct propagation and reflection, is instantaneously distributed over the whole specimen. In the conventional tensile test it is, in effect, assumed that, corresponding to each level of increasing, applied force or stress, there is an average, uniformly distributed strain that is the same as would be found if the force were held at the particular level indefinitely — that is, applied statically — and there were no creep effects. For this reason such tests are called "static" or "quasi-static." From measurements of the force at one end of the specimen, and of the overall elongation, at each instant or periodically during such a test, there is obtained a functional relationship between average stress and average strain: the stress-strain, or tenacity-strain curve (see Section 2.2).

4.1.2 <u>Longitudinal Impact.</u> When the velocities of impact become large, and especially when they approach the velocity at which strain is propagated in a test material, erroneous conclusions might be drawn from the assumption that stresses and strains are uniform along the specimen during the impact process. For a more realistic understanding of the process, one must take into consideration the fact, for instance, that at one instant the strain may be very great at one point in the specimen, and yet

---

*In the present discussion the term "elongation" will be used to designate increase in length in units of length (cm, in.), while "extension" and "strain" will have the usual connotation of proportional deformation.

nonexistent at another point. To obtain a more detailed picture,
let us analyze the elementary process.

<u>Mechanism of strain propagation.</u> Let it be supposed that the
end of a long rod, representing a textile fiber or yarn, is initially at rest at the position A in Figure 4.1.   If this end is
moved to the right by suitable means, at a constant (instantaneously attained) impact velocity v_0, at the same instant t_0, a condition of longitudinal strain (namely, the minute displacements of
atoms or molecules which represent strain) will start to move
from the end toward the left in the rod with a certain velocity.

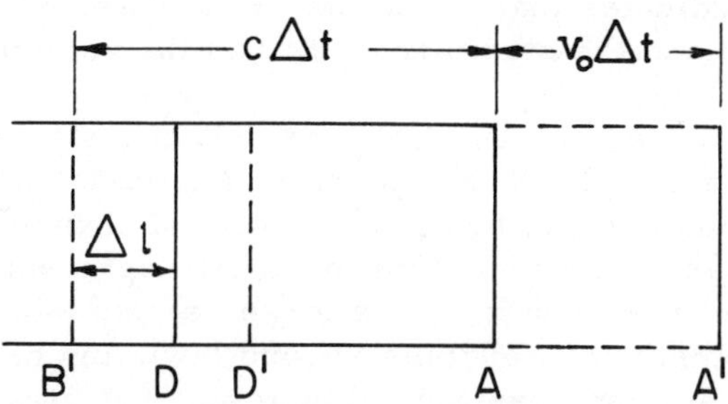

Fig. 4.1. Deformation and strain propagation in rod after
longitudinal impact.

In general, this velocity will vary as the strain increases; but
during a small time interval Δt, it may be considered to have
the constant value c. After the time interval Δt, the impacted
end will have travelled a distance $v_0 \Delta t$ and be at the position A'.
In the same time Δt, the strain front* will move a distance $c\Delta t$
to B'. Ahead of this front the elementary particles of the material will be, as yet, completely unaffected by the impact. At
time $t = t_0 + \Delta t$, the only portion of the bar in a state of strain is
the segment A'B'.

Evidently, an original length of rod AB' now has the length
A'B'; that is, an elongation A'A has occurred. The small strain
ϵ_0 occurring in the time increment Δt, is given by A'A/AB' and
thus

$$\epsilon_0 = \frac{v_0 \, \Delta t}{c \, \Delta t} = \frac{v_0}{c} \qquad (4.1)$$

where c is called the velocity of propagation of strain (or sometimes, velocity of the strain wave). The relationship given by
Equation 4.1 was established for tensile impact by Thomas Young
in 1807.[114]

---

*It is assumed that plane cross sections of the unstrained ma-
terial remain plane under strain, so that an initially plane strain
front advances as a plane.

Consider now any plane D (Figure 4.1) which was originally a distance $\Delta\ell$ from the plane that at time t is occupied by the front B'. If we assume homogeneous strain (which is implicit here), the plane D will have moved a distance $\epsilon_0 \Delta\ell$, to the position D', in the time required for the strain front to move the distance $\Delta\ell$. This time is given by $\Delta\ell/c$. Hence, during this interval the velocity of the plane D is

$$\epsilon_0 \frac{\Delta\ell}{\dfrac{\Delta\ell}{c}} = \epsilon_0 c = v_0 \tag{4.2}$$

This result indicates that the whole of the mass of the rod between the end and the strain front B' has the velocity v_0 at the instant $t = t_0 + \Delta t$.

The mass of the segment A'B' per unit cross section is $\rho c \, \Delta t$, where ρ is the density of the unstrained material. Hence the momentum at time t is $\rho c \, \Delta t \, v_0$, and the rate of change of momentum, $\rho c v_0$. The change of momentum in the AB' segment is produced by the force of the impact. This force is equal to the rate of change of momentum (Newton's second law), and hence, referred to unit cross section, has the value $\rho c v_0$. The strain ϵ_0 produced by the impact sets up a stress σ_0 sufficient to balance the applied force on unit cross section at every instant. Thus, $\sigma_0 = \rho c v_0$. But, from Equation 4.1, $v_0 = c\epsilon_0$, so that

$$\sigma_0 = \rho c^2 \epsilon_0 \tag{4.3}$$

In general, the stress set up in the material by the strain could be expected to be a function of this strain, as well as of other variables. The stress, therefore, is given, in terms of strain, by $\sigma_0 = (\partial\sigma/\partial\epsilon)\epsilon_0$, so that Equation 4.3 becomes

$$\frac{\partial\sigma}{\partial\epsilon} = \rho c^2 \tag{4.4}$$

From this, for the velocity of propagation of the strain front

$$c = \sqrt{\frac{1}{\rho}\frac{\partial\sigma}{\partial\epsilon}} \tag{4.5}$$

It can be seen from Equation 4.5 that c is not, in general, a constant, but depends on $\partial\sigma/\partial\epsilon$, the slope of the stress-strain curve. However, if the value of $\partial\sigma/\partial\epsilon$ is taken at all times as that corresponding to the strain front, where $\epsilon_0 = 0$, $\partial\sigma/\partial\epsilon$ will be the slope of the stress-strain curve at the origin, a unique value. What this means, then, is that the strain front, the plane B' in Figure 4.1, travels with a constant velocity. The trailing increments of strain, however, will in general travel with a different velocity; lower, if the stress-strain curve is concave-downward. In fact, this is the necessary condition to prevent the trailing increments of strain from overtaking the strain front — "shock-wave" phenomena.

If, at all levels of strain, $\partial\sigma/\partial\epsilon$ has the same value as it has at $\epsilon_0 = 0$, we have a linear stress-strain curve, implying that Hooke's law is obeyed. Then we may place $\partial\sigma/\partial\epsilon = E$ (Young's modulus) a constant up to the yield point of the material. Thus, Equation 4.5 becomes

$$c = \sqrt{\frac{E}{\rho}} \qquad\qquad (4.6)$$

which is the familiar expression for the velocity of propagation of a deformational wave (such as sound) through an elastic medium.

The relationship given by Equation 4.1 provides an estimate of the limiting velocity at which a free, unclamped yarn can be impacted longitudinally without immediate rupture at the point of impact. This limiting velocity evidently cannot be greater than $c\epsilon_b$, where ϵ_b is the breaking extension of the material. Interpreted physically, this means that the motion of the impacted end of the yarn cannot be transmitted to the rest of the yarn because the breaking strain is reached immediately on impact. As Smith and associates[97] point out, for example, a Fiberglas yarn, having $c = 3500$ m/sec, and ϵ_b = about 2%, would be likely to break immediately if impacted at a velocity of $3500 \times 0.02 = 70$ m/sec. A yarn of some other material having the same ϵ_b, but a higher c, would be set in motion by the impact, without rupture.

Reflection of strain front. Consider a yarn attached at both ends to rigid, but movable masses, and suppose one of these masses, the "head," is rapidly set in motion so as to give the attached end of the yarn an impact velocity.

The strain ϵ_0 produced at the end of the yarn, through the head mass, will travel backward in the specimen in the manner described in the preceding paragraphs, until it reaches the "tail" mass. Here the strain front will be reflected and propagate toward the head. The strain will be approximately fully reflected, so that at the tail immediately after the front is reflected, the strain will be doubled. Accompanying the propagated strain is a state of tension, roughly proportional to the strain (depending on how well the material obeys Hooke's law). This tension on reaching the tail has the effect of accelerating this mass.

The reflected strain front, on reaching the head will again be reflected. The accompanying tension here has the effect of further decelerating the head mass. Reflections will continue to occur at head and tail at time intervals of L/c, where L is the length of yarn between head and tail. At each reflection the local strain is increased by about $2\epsilon_0$, twice the initial strain, and the local tension is correspondingly increased. The process

is illustrated schematically in Figure 4.2, from Stone and associates.[104] Quoting them, with regard to this representation:
"Before impact the specimen is represented in A by a series of
adjoining circular springs.   After impact in B, the circular
spring next to the head is deformed into an ellipse and the strain
pulse is indicated by the adjacent cross-hatched circle and ellipse.   The arrow indicates the direction of strain propagation.
Sketches C and D show the arrival and reflection of the strain
pulse at the tail.   At this moment the tail is set in motion, and
the ellipse before reflection is deformed to one of greater eccentricity after reflection.   Sketch E indicates the arrival of
the reflected strain pulse at the head.   After reflection at the
head, the strain pulse is shown in sketch F as travelling toward
the tail about midway between the head and tail.   In G the strain

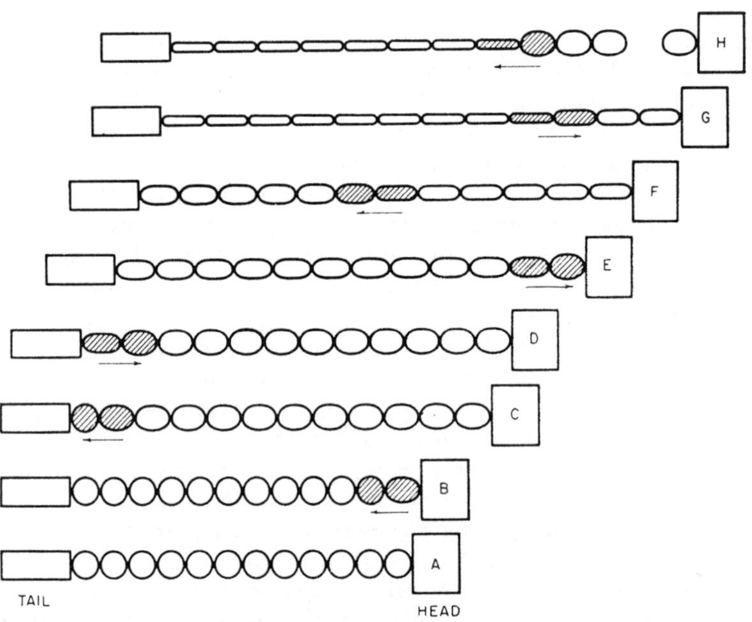

Fig. 4.2. Schematic diagram of strain propagation in linear
specimen, represented by adjoining circular springs, after
impact of head (toward right) at an instant between configurations A and B.[104]

pulse is shown as travelling toward the head after reflection at
the tail, and the high strain is indicated by the highly eccentric
ellipses.   After the next reflection at the head, it is assumed
that the rupture strain is attained, and the ruptured specimen is
shown in H.   After rupture the strain is released.   This condition is represented in H by the reconversion (recovery) of the

ellipses of high eccentricity to ellipses of low eccentricity or to
circles for complete recovery.   The propagation of the strain-
release pulse is indicated by the arrow. The ruptured end will
snap toward the tail with a very high velocity, and the momentum acquired by the specimen when the strain-release pulse
reaches the tail, may be sufficient to stop the tail mass or to reverse its motion."

The corresponding process that occurs when the tail end is
immovably clamped is illustrated in more quantitative fashion
in Figure 4.3, taken from the article of Smith and associates.[96]
Here the length of the filament or yarn is taken as defining a
negative x-axis, with the head end at $x = 0$, and the fixed end at
$x = -L$.   As was assumed in the above discussion of the mechanism of strain propagation, the head end instantaneously, at

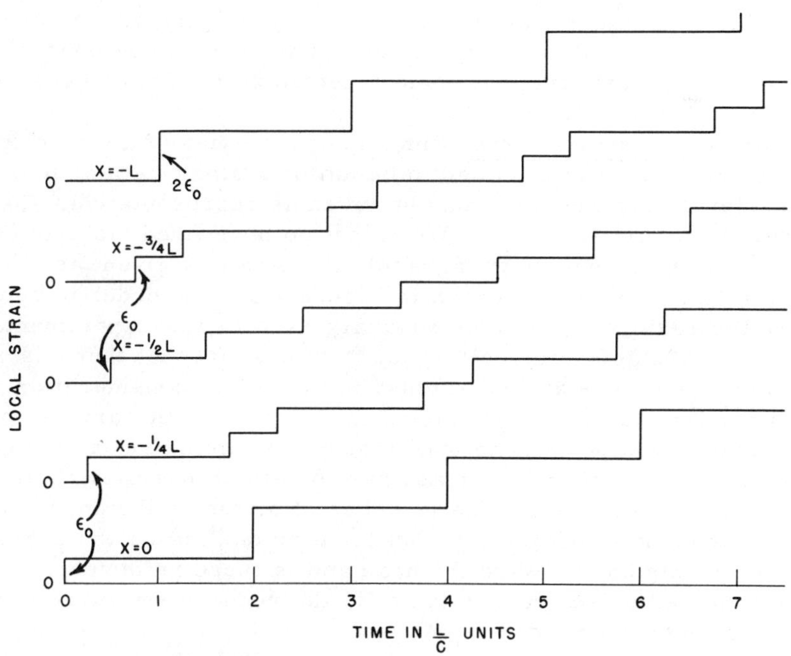

Fig. 4.3. Local strain as a function of time, in a filament
clamped at $x = -L$ and rapidly elongated at $x = 0$.[96]

time $t = 0$, acquires a velocity v_0 in the positive direction.   A
strain pulse of magnitude $\epsilon_0 = v_0/c$ will travel along the yarn in
the negative direction with velocity c.   Reflections will occur
alternately at the fixed end and at the head, as described above.
In Figure 4.3 are plotted theoretical graphs of the strain at various points (local strain) along the yarn ($x = 0$, $L/4$, $L/2$, and so forth),

as functions of time. Since the position of the pulse at any instant depends on its velocity c and the distance L which it must travel before reflection, the time (in the general case) must be expressed as a multiple of L/c.

The strain pulse of magnitude ϵ_0 arriving at either end is itself fully reflected, since the end masses are assumed to be so large and rigid as to absorb none of the strain. Thus, the strain at either end is increased by the amount $2\epsilon_0$ on each reflection. While the strain at each point increases with time, it can be readily seen that the strain is not uniform at all times along the length of the yarn. Thus, for example, an instant before $t = 3L/c$, the strain is $3\epsilon_0$ at $x = 0$, $-L/4$, $-L/2$, and $-3L/4$, and is $2\epsilon_0$ at $x = -L$. At a slightly later instant, however, at (say) $t = 27L/8c$, the strain distribution has altered considerably: The local strain is still $3\epsilon_0$ at $x = 0$, $-L/4$, and $-L/2$, but is $4\epsilon_0$ at $x = -3L/4$ and $-L$. Since the strain in the yarn at any instant varies from point to point, it is clear that one cannot deal precisely with a single strain for the whole yarn specimen. The strain to be considered must be the local strain at some selected point. This local strain itself varies with time.

The stress-strain curve. Since the properties of the yarn specimen are implicitly assumed to be uniform along its length, the behavior at any one point may be taken as representative. Consider the strain at $x = -L$. When that point is fixed (as was assumed in the foregoing paragraph), the strain is given, as a function of time, by the top graph in Figure 4.3. In order to compare the results of this type of analysis with the experimental results of Schiefer and colleagues,[60, 104] the fact that there is a movable tail mass at $x = -L$ must be taken into consideration. As was brought out in the discussion of Figure 4.2, the arrival of the first strain pulse at the tail mass sets this mass in motion. This movement of the tail mass has the effect of relieving the strain at $x = -L$, so that the increase of strain by the amount $2\epsilon_0$ does not remain indefinitely, but tends to decline as the strain pulse travels back toward the head and is there reflected. When the reflected pulse reaches the tail, the reduced strain is again given an increment $2\epsilon_0$.

This process continues until the velocity of the repeatedly accelerated tail mass reaches such a value that the strain at $x = -L$ is decreased between reflections at that point by amounts increasingly greater than $2\epsilon_0$. Ultimately, in the ideal case, all strain in the specimen will disappear. With a movable tail mass, therefore, the theoretical local strain-time curve for $x = -L$ in Figure 4.3 is distorted to the shape of the graph in Figure 4.4, in which strain is expressed in units of ϵ_0. The curve was calculated by Smith and associates from the solutions* of the classical wave equatio

---

*The applicable solutions are given in an appendix of the article of Smith and associates.[96]

$$\frac{\partial^2 u}{\partial t^2} = c^2 \frac{\partial^2 u}{\partial x^2} \tag{4.7}$$

where u = the displacement of a cross section at x from its position in the unstrained state of the specimen. The displacement u is related to the strain by the equation $\epsilon = \partial u/\partial x$. The analysis involving Equation 4.7 is beyond the scope of this monograph, but may be found in the original article.[96]

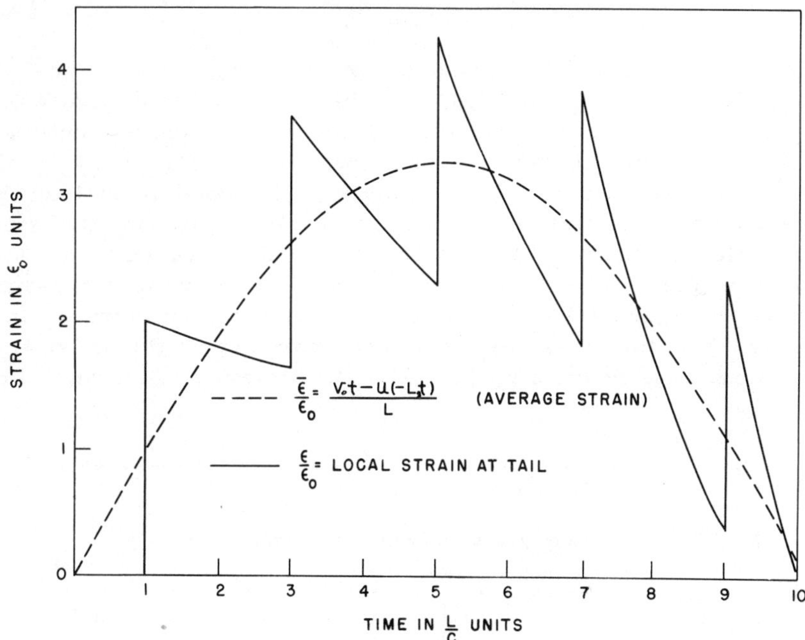

Fig. 4.4. Strain as a function of time, in a filament having a mass at one end (tail) and being rapidly elongated at the other end (head).[96]

The experimental technique of Schiefer and colleagues provides a high-speed photographic record of the motion of the tail mass, from which a velocity-time curve can be obtained.   From the slope of this curve and the mass of the tail, the force exerted on it by the strained yarn can be computed, as has been indicated in Chapter 3. Thus, the local force or stress at the tail can be obtained as a function of time. From measurements on the record of the motions of both the head and tail masses the average strain in the specimen can evidently be obtained as a function of time. Correlating these data at corresponding times yields a local stress-average strain curve.   To develop the comparable theoretical curve it is necessary to obtain representations of average strain, as well as of local stress at the tail, as functions of time.

In the case of a movable tail mass, the average over-all strain $\bar{\epsilon}$ in the yarn at any time t is given by the initial strain $\epsilon_0 = v_0/c = v_0 t/L$ (where t is expressed in multiples of L/c), less the decrease in strain due to the displacement of the tail end.   The displacement of the tail end is given by the quantity u, which, in general, is a function of the position x, and the time t. For the present purposes, u is evaluated at x = -L. Smith and colleagues give the relationship

$$\bar{\epsilon} = \frac{v_0 t - u(-L, t)}{L} \tag{4.8}$$

and have calculated this average strain, using the solutions of Equation 4.7 in Equation 4.8, with the result shown in Figure 4.4.

The solutions of Equation 4.7 obtained in the above-mentioned analysis assume the validity of Hooke's law. For many textile materials, the actual behavior under high-speed impact sufficiently approximates this condition, thereby justifying its assumption. Hence, the local stress at the tail can be taken as proportional to the local strain at that point.   The local strain-time curve in Figure 4.4 can then be taken as the local stress-time curve, with the stress expressed in arbitrary units, or conventional units if Young's modulus for the material is known.   The

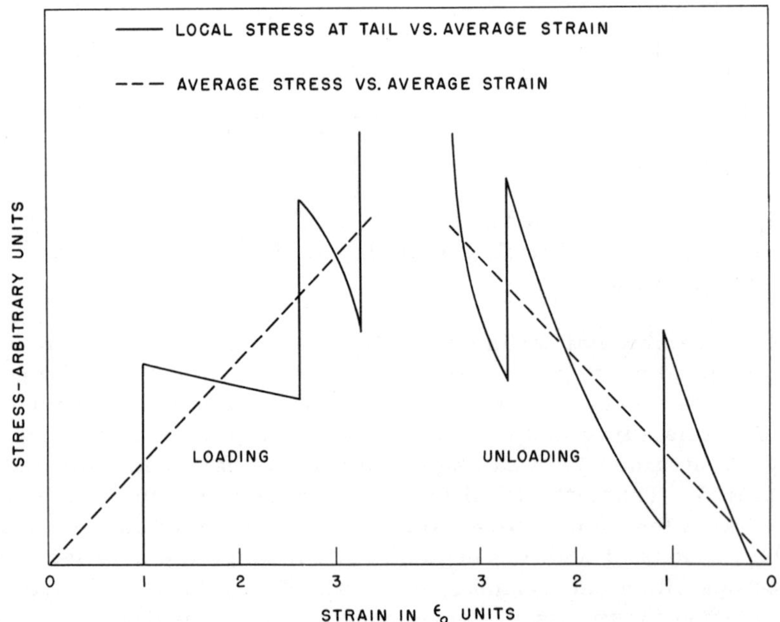

Fig. 4.5. Local stress at tail, as a function of average strain in a filament rapidly elongated at head.[96]

local stress at various times can then be correlated with the average strain. The plot of such a relationship is shown in Figure 4.5, for both the loading and unloading of the yarn.

In these theoretical graphs, now, the quantities plotted (stress at the tail and average strain) are the same as those obtained experimentally by Stone, Schiefer, and Fox.[104] The loading and unloading curves they obtained for a nylon yarn are shown in Figure 4.6. The delay in the rise of stress at the tail until the arrival of the strain pulse, and the subsequent stepwise increase in stress with increasing average strain, which are seen in Figure 4.5, are found experimentally. Presumably, this behavior accounts for the regularity with which the experimental points for the loading half-cycle in Figure 4.6 are scattered about the average curve. It can be seen that the dashed curve drawn through the experimental points has a sinusoidal form. Analysis of the data for the nylon yarn by Schiefer and colleagues, indicated that this sinusoidal effect has a period of about 6×10^{-4} sec. Since this represents the time required for the strain pulse to travel from the tail to the head and back again, the velocity of strain propagation can be calculated. Such a calculation yielded for the nylon yarn (0.65 m long) a propagation velocity of about 2200 m/sec.

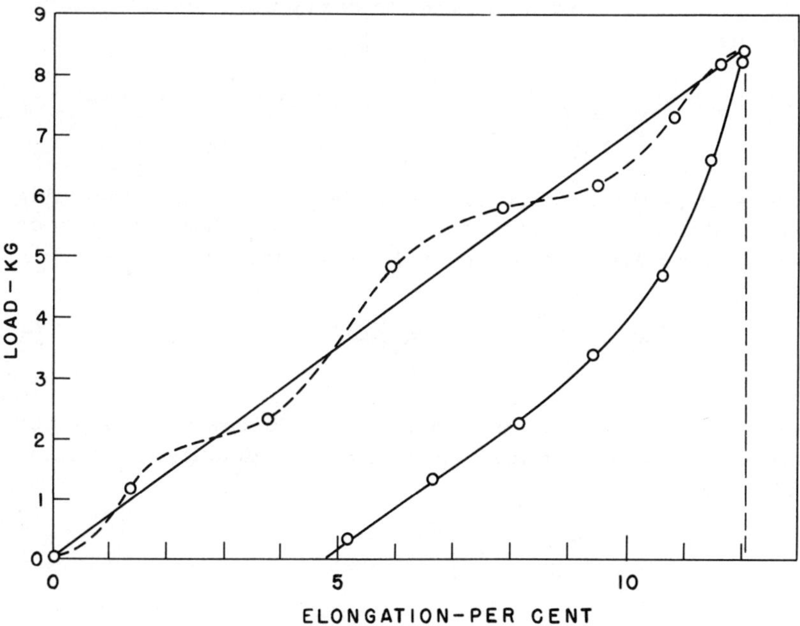

Fig. 4.6. Force (load) at tail as a function of average extension (elongation), for the impact loading and unloading of a nylon 66 yarn.[104]

Breaking energy density. While breaking stress or tenacity
is an important criterion of the impact behavior of a textile material in certain applications, and breaking strain or extension
is important in other applications, a more general parameter for
characterizing impact resistance is breaking energy density. This
quantity is the work per unit mass of specimen required to strain
the specimen until rupture occurs. The breaking energy density
is proportional to the area under the stress-strain curve. In the
quasi-static test, for which the stress-strain relationship is likely
to be accurately known, the breaking energy can be obtained in
this way. In impact tests, however, the stress-strain curve appropriate to the high-speed conditions may not be known, or not
known with sufficient certainty to allow the use of this technique.

Generally, in impact testing, breaking energies can be found
directly by measuring the kinetic energy lost to the textile specimen by the impacting body. In the falling-pendulum type of impact experiment, for example, this loss of kinetic energy has been
obtained by measurement of the height to which the pendulum,
after impact, swings from a fixed initial position, as has been
discussed in Chapter 3.

If a body having unit length and cross section is given an infinitesimal extension $\Delta\epsilon$ by a tensile stress σ, the work done
on it will be the product of the force by the displacement $\sigma \Delta\epsilon$.
The amount of work necessary to carry the body to rupture will
then be

$$w_b = \int_0^{\epsilon_r} \sigma \, d\epsilon \tag{4.9}$$

where ϵ_r is the rupture strain in the general case, including the
quasi-static. If the material obeys Hooke's law, the stress is
given by $\sigma = E\epsilon$, and Equation 4.9 becomes

$$w_b = \int_0^{\epsilon_r} E\epsilon \, d\epsilon = \tfrac{1}{2} E\epsilon^2 \tag{4.10}$$

Since w_b refers to a unit cube, it is the breaking energy per unit
volume. Hence, for the breaking energy density, as defined earlier, we have

$$W_b = \frac{w_b}{\rho} = \frac{1}{\rho} \int_0^{\epsilon_r} \sigma \, d\epsilon \tag{4.11}$$

or for Hookean behavior

$$W_b = \frac{E\epsilon_r^2}{2\rho} \tag{4.12}$$

While Equations 4.11 and 4.12 are definitive, analytical expressions for an important parameter, they are not very useful
for calculating W_b, since the direct measurement of ϵ_r is not
feasible, nor is the function σ known <u>a priori</u> with much accu-
racy. The device of measuring the kinetic-energy loss, to ar-
rive at the breaking energy of a yarn sample, generally cannot
be used with the high-speed experimental techniques, such as that
of Schiefer and colleagues, because the impacting mechanism has,
relatively, such enormous kinetic energy that the loss is imper-
ceptible. Approaching the problem with a view to the dynamics
of the system of yarn specimen and head and tail masses, Smith
and colleagues[97] have derived an expression for breaking energy
density involving only experimentally measurable quantities. The
following treatment substantially follows that of these authors.

In the system to be considered the yarn is taken to have mass
m, the head mass hm, and the tail mass nm. Half of the mass
of the yarn is assumed to be concentrated in the head and half in
the tail. Strain propagation effects are neglected. When the
head mass is impacted longitudinally, it will be given an initial
velocity v_0. The corresponding initial momentum of the head is
$(h + \frac{1}{2})mv_0$. When the whole system is in motion its center of mass
has a velocity v_m, so that its momentum is $(h + n + 1)mv_m$. Because of the conservation of momentum, these two momenta
must be equal, so that we can write:

$$v_m = \frac{h + \frac{1}{2}}{h + n + 1} v_0 \qquad (4.13)$$

The effect of the impact will be to elongate the yarn. The
stress accompanying this elongation tends to decrease the velocity of the head and increase that of the tail, so that these velocities tend to become equal. At the instant this occurs these
velocities are also equal to v_m, and the specimen attains its maximum elongation.[60,97] If this is equal to the breaking elongation
of the material, it is at this instant that the specimen will rupture. The energy available to cause rupture in this system is
the kinetic energy lost by the decelerating head mass, minus the
kinetic energy gained by the tail mass. The yarn must be supplied with just the amount of energy necessary to cause rupture.
An excess would be wasted on extraneous effects, and lead to
spurious estimates of the breaking energy. The necessary
minimum energy is supplied when, for a given set of head and
tail masses, the head is given a certain sufficiently high, initial
velocity v_{hn}. This can be found experimentally by determining,
in a series of trials on a number of specimens of a sample, the
minimum velocity at which breakage can be made to occur.

The initial kinetic energy of the head mass will then be

$$\tfrac{1}{2} m(h + \tfrac{1}{2})v_{hn}^{\,2}$$

When rupture occurs, the head mass, as indicated above, will have velocity v_m, given by Equation 4.13, so that its kinetic energy at this instant is

$$\tfrac{1}{2} m(h + \tfrac{1}{2}) \left[\frac{(h + \tfrac{1}{2})^2}{(h + n + 1)^2} \right] v_{hn}^{\,2}$$

Hence the energy lost by the head mass will be

$$\tfrac{1}{2} m(h + \tfrac{1}{2}) \left[1 - \frac{(h + \tfrac{1}{2})^2}{(h + n + 1)^2} \right] v_{hn}^{\,2}$$

The kinetic energy gained by the tail mass, starting from rest, will be

$$\tfrac{1}{2} m(n + \tfrac{1}{2}) v_m^{\,2} = \tfrac{1}{2} m(n + \tfrac{1}{2}) \left[\frac{(h + \tfrac{1}{2})^2}{(h + n + 1)^2} \right] v_{hn}^{\,2}$$

(based on Equation 4.13)

Finally, the breaking energy, the difference between the latter two expressions, is found, after some algebraic manipulations, to be

$$\tfrac{1}{2} m \frac{(h + \tfrac{1}{2})(n + \tfrac{1}{2})}{h + n + 1} v_{hn}^{\,2}$$

Dividing this quantity by the mass of the yarn m gives the work done on unit mass, or breaking energy density,

$$W_{b,hn} = \tfrac{1}{2} \frac{(h + \tfrac{1}{2})(n + \tfrac{1}{2})}{h + n + 1} v_{hn}^{\,2} \tag{4.14}$$

where the subscripts h and n indicate that the energy density has been determined at the breaking impact velocity corresponding to head and tail masses hm and nm, respectively.

Limiting breaking velocity. If $W_{b,hn}$ for a particular material is taken to be a constant, independent of the impact velocity, we can identify it with W_b, mentioned previously. Evidently, then, v_{hn} will have to be increased to cause rupture if either the head or tail mass is reduced. Suppose that the tail mass is reduced to zero. In this case the mass of the yarn can be associated with the head only, so that the factor $(h + \tfrac{1}{2})$ becomes $(h + 1)$ and Equation 4.14 reduces to

$$W_b = \tfrac{1}{2} \frac{h + 1}{h + 1} v_{ho}^{\,2} \tag{4.15}$$

Since the terms containing h cancel out, v_{ho} is evidently independent of this experimental parameter, and can be taken as a characteristic of the material. Designating this as v_b, we thus have

$$W_b = \tfrac{1}{2} v_b^2$$

or

$$v_b = \sqrt{2W_b} \tag{4.16}$$

McCrackin and colleagues[60] have called v_b the limiting break-ing velocity. This value represents the maximum longitudinal impact velocity that a particular textile material can possibly withstand without rupture. The velocity v_b corresponds to that at which rupture strain ϵ_b is reached immediately on impact (discussed earlier in this chapter). This can be readily seen if we suppose a yarn of Hookean material to be impacted longitu-dinally so as to cause immediate rupture.   Then, from Equa-tions 4.1 and 4.6

$$\epsilon_b^{\,2} = \frac{v_i^{\,2}}{c^2} = \frac{\rho v_i^{\,2}}{E} \tag{4.17}$$

where v_i is the immediate-rupture velocity. The rupture strain, designated ϵ_r in Equation 4.9 is now ϵ_b. Hence, by Equations 4.12 and 4.16:

$$v_i = \sqrt{\frac{E}{\rho}}\,\epsilon_b = \sqrt{2W_b} = v_b \tag{4.18}$$

It should be noted, as a matter having practical implications, that this upper limit on longitudinal impact velocity cannot be raised by increasing the size or massiveness of the yarn or tex-tile structure into which the material is formed.

From Equations 4.11 and 4.16 we can write for the limiting breaking velocity,

$$v_b = \sqrt{\frac{2}{\rho} \int_0^{\epsilon_b} \sigma \, d\epsilon} \tag{4.19}$$

Through considerations of the propagation of what they term "plastic" strain in long rodlike solids, von Karman and Duwez[115] have derived an expression for a critical velocity v_c, above which instantaneous rupture on impact will occur. They obtain

$$v_c = \int_0^{\epsilon_c} \sqrt{\frac{1}{\rho}\frac{\partial \sigma}{\partial \epsilon}} \, d\epsilon \tag{4.20}$$

where ϵ_c is the maximum strain attained at impact velocity v_c. Since their theory leads to the conclusion that rupture is likely to occur when the strain reaches this point, ϵ_c can be taken equal to ϵ_b.

As Smith and associates[97] point out, v_c depends on the shape of the stress-strain curve, because of the presence of the slope, $\partial\sigma/\partial\epsilon$, in the expression on the right-hand side of Equation 4.20, whereas v_b depends on the area under the stress-strain curve, the integral $\int_0^{\epsilon_b} \sigma \, d\epsilon$. In general, for materials having stress-strain curves that are concave downward, v_c is less than v_b, by varying amounts. For materials obeying Hooke's law, v_b and v_c are the same, since the expressions on the right-hand sides of Equations 4.19 and 4.20 both reduce to $\sqrt{E/\rho}\;\epsilon_b$.   The von Karman theory applies strictly only to materials having stress-strain curves that are not concave upward, and that do not change in shape with rates of straining. Despite the fact that the stress-strain curves for most textile fibers and yarns have concave-upward segments, and are strain-rate dependent, the character of most of them at high strain rates is approximately Hookean, or predominantly concave-downward.*   Hence, the concepts underlying the principle of critical velocity can be, and have been, fruitfully applied in the analysis of the behavior of textiles under high-speed impact.

    4.1.3 <u>Transverse Impact.</u> When a linear structure as represented by a textile yarn, is struck transversely by a rigid object, the former is deflected in the direction of motion of the striking object or missile. A longitudinal strain is set up in the yarn at the point of impact. This strain is propagated outward along the yarn in both directions, in the manner described in the foregoing sections for strain resulting from longitudinal impact. As the strain fronts propagate outward, material of the yarn behind the fronts moves inward, as the yarn tends to accommodate itself to the new configuration.

This new configuration consists of a V-shape with the apex at the point of impact. As the impacting missile and the apex move forward, this V-shaped segment spreads outward along the undeflected yarn in both directions. This transverse motion of the yarn, generally called the <u>transverse wave,</u> is propagated with a certain velocity, which is always lower than that of the concurrent strain propagation. It will be of interest to consider the simple theory of transverse impact, and to establish some relationships between striking velocity, transverse-wave velocity, tension in the specimen, and so forth.

---

   * When the stress-strain or tenacity-extension curve of a textile has a concave-upward segment, shock-wave propagation results from impact. In this phenomena, the material ahead of the shock-wave front is completely unstrained, while that in the wake of the front has a single uniform strain.   Smith and colleagues have recently discussed this situation, and have shown that, in the region where the curve is concave-upward, the behavior is as if this were a straight line.[92]

<u>Transverse-wave velocity.</u> Consider a yarn lying in a horizontal plane, and suppose it to be impacted transversely at time t_0 by a missile having a striking velocity v_s. The motion of the missile being represented as vertical, the configuration of the yarn at some time t after impact (and before any strain reflection occurs) will be that shown by the solid line in Figure 4.7.[*]

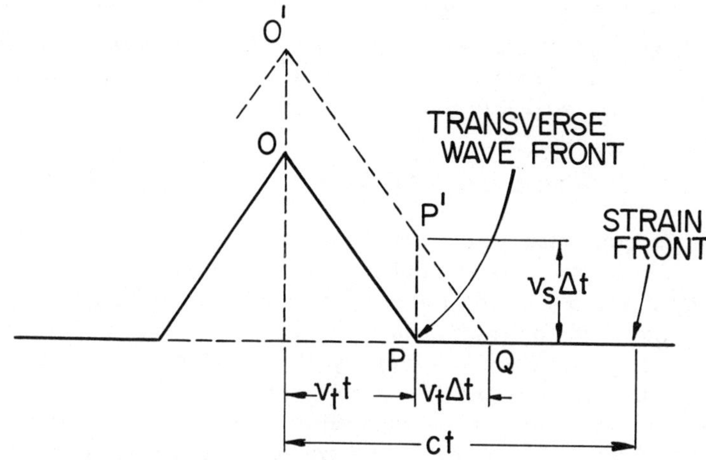

Fig. 4.7. Configuration of yarn, and strain propagation after transverse impact (vertically).

The configuration being always symmetrical about the apex, we need concern ourselves with the behavior of the yarn in only one half of the inverted V, say the right-hand side in Figure 4.7. At time t, this half is in the position OP. At some slightly later time $t + \Delta t$, the yarn will have moved to O'P'Q. The point on the yarn originally at P will be displaced parallel to the trajectory of the missile, to the position P', in the time Δt. The distance PP' = OO'. It is assumed that the energy of the missile is sufficiently large that, compared to v_s, the loss of velocity, not only

---

[*] In the present treatment, the effect of air resistance, which produces a drag on the transversely moving yarn, is neglected. The effect of this drag is to bow downward slightly the yarn segments represented by OP in Figure 4.7. McCrackin[59] has recently presented a theoretical treatment of transverse impact (outside the scope of this monograph) taking into account the influence of air drag. He shows that the differences introduced by considering the drag are small. Applying his results to experimental data on a nylon yarn, he calculated a strain of 2.03% at one stage in the impact process, neglecting air drag; taking drag into account this strain was found to be 2.12%.

in the interval Δt, but in the whole transverse deformation process is negligible. Hence we may place $OO' = v_s \Delta t$, so that we have $PP' = v_s \Delta t$. If we take v_t as the velocity with which the transverse-wave front moves from P to Q, the distance $PQ = v_t \Delta t$.

Let M = the mass per unit length (linear density) of the yarn in the unstrained state. Then the mass per unit length of the yarn having a total strain ϵ will be $M/(1 + \epsilon)$. If ϵ is the strain in the segment P'Q, then the mass of this segment will be $[M/(1+\epsilon)]\sqrt{v_s^2 + v_t^2}\,\Delta t$, where $\sqrt{v_s^2 + v_t^2}\,\Delta t$ is evidently the length P'Q. The transverse momentum of this segment, travelling at the velocity v_s, will then be

$$p = \frac{M\,v_s}{1 + \epsilon}\sqrt{v_s^2 + v_t^2}\,\Delta t \tag{4.21}$$

All of this momentum represents a gain during the interval Δt, since the segment was at rest at PQ at time t.

By the principle of the conservation of momentum this gain must be equal to the momentum lost by the missile during the interval Δt. This loss is given by the product of the rate of change of momentum and the time interval. But this rate of change of momentum is the retarding force acting on the missile. This force, in turn, is the transverse component of the tension or stress σ in the strained yarn, multiplied by the cross-sectional area A of the latter. According to the resolution of vectorial quantities, the transverse-force component bears to the tensile force σA in the yarn the ratio of the lengths PP'/P'Q, that is, $v_s \Delta t/\sqrt{v_s^2 + v_t^2}\,\Delta t$. Thus, the retarding force acting on the missile is given by

$$f = \frac{v_s \sigma A}{\sqrt{v_s^2 + v_t^2}} \tag{4.22}$$

Equating the gain of momentum of the yarn p, given by Equation 4.21, to the loss of momentum of the missile, $f \Delta t$, we have

$$\frac{Mv_s}{1 + \epsilon}\sqrt{v_s^2 + v_t^2}\,\Delta t = \frac{v_s \sigma A\,\Delta t}{\sqrt{v_s^2 + v_t^2}} \tag{4.23}$$

Simplifying this equation, and noting that $M/A = \rho$, the (volume) density of the unstrained yarn, we have:

$$\frac{\rho}{1 + \epsilon}\left(v_s^2 + v_t^2\right) = \sigma \tag{4.24}$$

Rearranging Equation 4.24 gives us an expression for the transverse-wave velocity:

$$v_t = \sqrt{\frac{\sigma(1 + \epsilon)}{\rho} - v_s^2} \tag{4.25}$$

The velocity v_t, as derived above, and expressed by Equation 4.25, is relative to a coordinate system at rest with respect to the laboratory. Since the unstrained portion of the yarn (ahead of the strain front) is at rest with respect to the laboratory, v_t is relative also to this portion of the yarn.

An expression that is sometimes useful is that for the transverse-wave velocity relative to the strained segment of yarn (behind the strain front), which flows inward to form the inverted V. Consider a particle in the strained yarn. As time goes on, the particle will move inward with increasing velocity, and the strain in its neighborhood will increase, until so-called plastic strain is reached. Under this condition the inward-flow velocity becomes constant, at v_f, as does also the strain, at ϵ_p. Relative to the particle, in the expanding coordinate system of the strained state, the transverse wave is advancing on the particle with velocity

$$v_{t'} = \frac{v_t + v_f}{1 + \epsilon_p} \tag{4.26}$$

From other considerations, Smith and associates[98] have shown that

$$v_{t'} = \sqrt{\frac{\sigma}{1 + \epsilon_p}} \tag{4.27}$$

From what has been said about the constancy of the strain after the plastic level is reached, it is evident that $\epsilon = \epsilon_p$. This strain is not altered by the passage of the transverse front. Thus, there is a uniform distribution of strain in the transverse wave, except in the rare cases of some materials in which the transverse front propagates faster than the plastic-strain front.

The stress-strain curve. Smith and associates[100] have developed procedures based on the foregoing theory, for obtaining the stress-strain curve of a yarn sample from observations of transverse-impact experiments. The method to be described here is applicable when the striking velocity is sufficiently low that the difference between local strain ϵ_p at the transverse front and average strain $\bar{\epsilon}$ is negligible.

From measurements on high-speed photographs of the yarn at successive times after impact, the lengths L_1 of the transverse wave (OP, O'Q, and so forth in Figure 4.7) and the lengths L_2 of the undisplaced yarn from the transverse front (points P, Q, and so forth) to the end of the yarn, are obtained. The average strain $\bar{\epsilon}$ at each time is then computed using the relation

$$\bar{\epsilon} = \frac{L_1 + L_2 - L}{L} \tag{4.28}$$

where L equals the initial length of the unstrained yarn from the

point of impact to the end. The successive values, at known
time intervals, of L expressed in terms of the expanding coordinate system, provide measures of $v_{t'}$. Then the tension or
stress is computed by means of the relation

$$\sigma = v_{t'}^{2}(1 + \overline{\epsilon})\rho \tag{4.29}$$

which is derived from Equation 4.27, with $\overline{\epsilon}$ replacing ϵ_p. The
stress-strain curve is obtained by plotting values of $\overline{\epsilon}$ against
those of σ at corresponding times. A typical such curve for a
high-tenacity nylon yarn is shown in Figure 4.8 for a rate of
straining of 5000 %/sec (300,000 %/min).* Compared with this
curve are those obtained at three quasi-static rates of loading.
  The foregoing analysis, applied to nylon data, shows that increasing the rate of straining tends to displace the stress-strain
curve to the region of higher stress for a given strain. Schiefer
and colleagues,[87] using the same rapid impact techniques, and

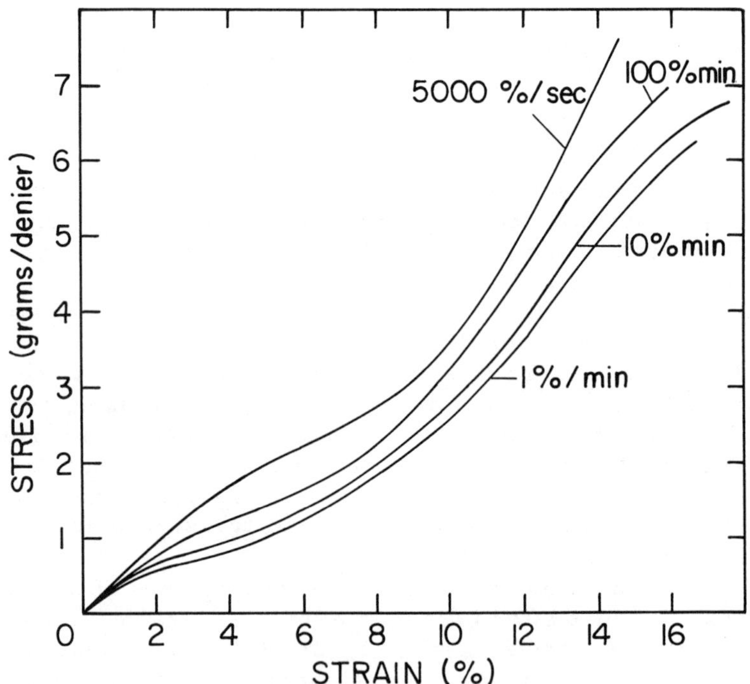

Fig. 4.8. Tenacity (stress)-strain curves of high-tenacity
nylon 66 yarn at various strain rates. [After Reference 100.]

---

  * This curve is in excellent agreement with one obtained on
the same material by longitudinal impact at an initial straining rate of 315,000 %/min.[87]

before them, Peirce[76] and others,[51,63] using lower speeds, have found the same effect on a wide variety of fiber types.

Distributions of strain after impact. Petterson, Stewart, and associates,[78,80] with their ballistic techniques which produce initial rates of straining in the order $10^6\%$/sec, have thrown further light on the influence of the rate of straining on the stress-strain behavior of textile materials. By placing ink tick marks on the test yarn at regular intervals and employing high-speed photography, these workers have obtained full records of local strain, before and at various times after transverse impact. In Figure 4.9 is shown an idealized drawing of a triple-flash photograph of such a marked yarn in the neighborhood of the transverse wave, obtained as described in Chapter 3. As is brought out there, the experimental technique is such that the images of the yarn before impact, and at the three different times after impact are slightly displaced horizontally with respect to each other. It will be seen that points in the transverse wave do actually travel substantially parallel to the trajectory of the missile as was postulated in connection with the discussion of Figure 4.7.

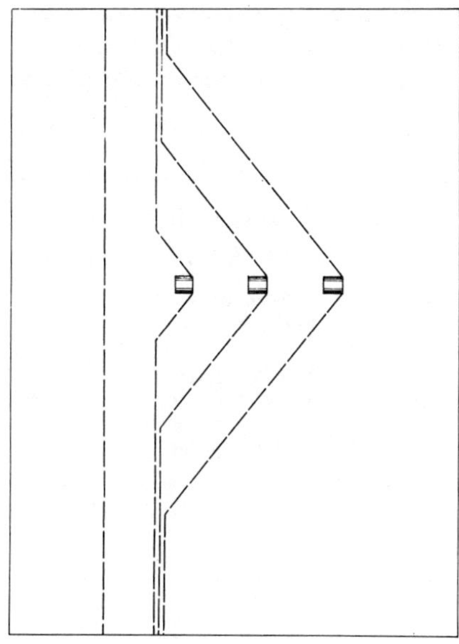

Fig. 4.9. Idealized drawing of successive images of yarn impacted transversely, as obtained by triple-flash technique.[39,40]

By means of this technique it has been possible to obtain the local strain in the yarn as a function of distance x from the point of impact (the axial distribution of strain) at selected times t after impact, and at various transverse striking velocities. The measurements were made in sufficiently short times so that no reflection effects enter. A typical set of curves for a nylon sample, obtained from the records of four shots at an average striking velocity of 235 m/sec is shown in Figure 4.10. Two important observations were made on these data. First, it was found that, for a particular striking velocity, the average of the velocity of propagation x/t of a selected strain (such as ϵ_i in Figure 4.10) during increasingly longer intervals after impact was substantially constant. This finding implied that a constant propagation velocity is associated with this strain ϵ_i.   A different velocity was found to be associated with each level of strain, the lower strains having the higher velocities. Second, it was found that while the propagation velocity of a particular strain generally increases with striking velocity, the velocity of the strain front (where the strain is infinitesimally small) has constant value for all striking velocities. These experimental findings correspond to the concept of the "elastic wave," advancing with a certain velocity, followed by a slower "plastic wave," introduced by von Karman[115] and further interpreted by Smith and associates.[98]

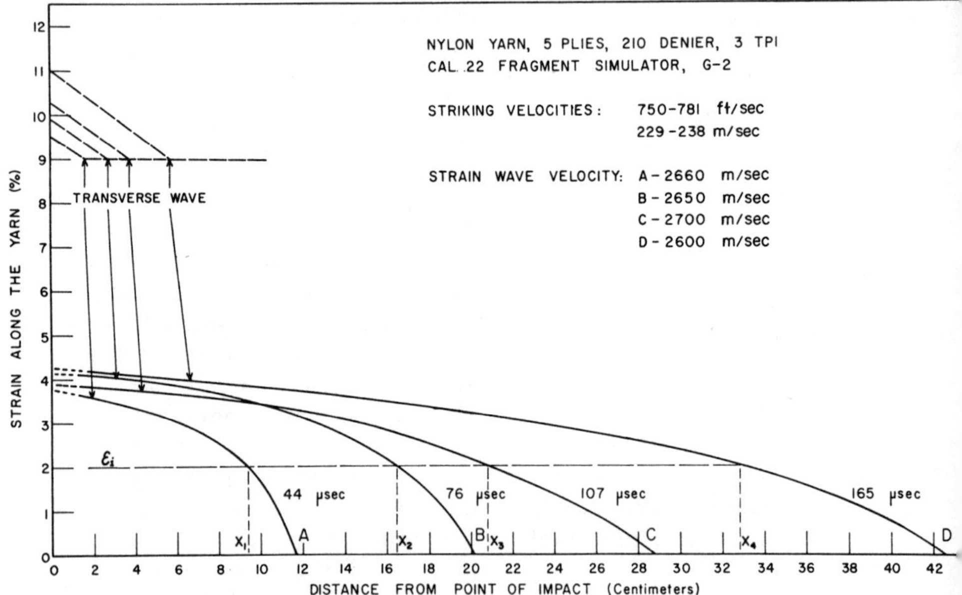

Fig. 4.10. Strain as a function of position along yarn, at four different times after transverse impact.[80]

In Figure 4.11 are summarized results on a nylon yarn at four average transverse striking velocities, showing the effects noted earlier. It would be desirable to obtain the relationship between strain and strain velocity over the greatest possible range. The upper limit of the range in which the above-described technique can be applied is evidently the relationship corresponding to the transverse critical velocity (T.C.V.), at which, in theory, the specimen breaks immediately on impact. At this and higher velocities no record translatable into ϵ and x/t values is obtainable.

The form of the ϵ-vs.-x/t curve corresponding to the trans-

verse critical velocity can be calculated if a relationship between ϵ and x/t, based on the experimental data for lower velocities, is known.* Petterson and Stewart[78] found such an empirical relationship for each striking velocity:

$$\epsilon = \frac{\epsilon_m}{1 - e^{-k}} \left[1 - e^{-k[1-(x/ct)]} \right] \qquad (4.30)$$

where ϵ_m = the maximum strain in the yarn, occurring in the
neighborhood of the projectile;† k = a dimensionless constant
characteristic of the particular striking velocity; and c = velocity
of propagation of the elastic strain front (the value 2700 m/sec
in Figure 4.11, for instance).

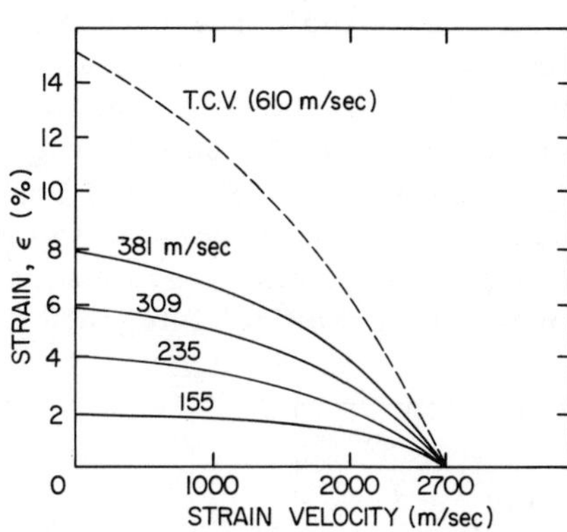

Fig. 4.11. Strain in yarn as a function of strain velocity x/t, for four different transverse impact velocities. [After Reference 78.]

From the measurable values of ϵ, ϵ_m, x/t, and c, values of k corresponding to various transverse striking velocities can be calculated by means of Equation 4.30. Then, considering ϵ_m and k as functions of striking velocity, their values can be obtained graphically at the known transverse critical velocity. Substituting these values in Equation 4.30, one can then calculate ϵ for various values of the strain-propagation velocity x/t, and thus obtain the form of the curve at the transverse critical velocity. The function obtained by Petterson and Stewart by the nylon yarn is shown as a dashed line in Figure 4.11.

In order to obtain the stress-strain curve, these authors make use of Laplace's wave equation given earlier (Equation 4.7) in the form

$$\frac{\partial^2 u}{\partial t^2} = \frac{E_i}{\rho} \frac{\partial^2 u}{\partial x^2} \qquad (4.31)$$

---

* Being based in part, on data for lower velocities, the calculation is substantially an extrapolation of these data, and has the uncertainties inherent in such a practice.

† The strain ϵ_m will be recognized as being the ϵ_p discussed earlier in this section. Petterson and Stewart represent ϵ_m as having zero propagation velocity, rather than a finite constant velocity. Their data, however, probably do not preclude the latter interpretation.

where E_i is the instantaneous modulus corresponding to the strain ϵ_i at time t_i. Since $\epsilon = \partial u/\partial x$, we can obtain an expression for u by integration of Equation 4.30 with respect to x, and application of the boundary condition: displacement $u = 0$ at $x = ct$. This operation leads to

$$u = \frac{\epsilon_0}{1 - e^{-k}} \left\{ \frac{ct}{k} \left[1 - e^{-k[1-(x/ct)]} \right] - t\left(c - \frac{x}{t}\right) \right\} \qquad (4.32)$$

Taking the second partial derivatives with respect to t and x, separately, and substituting the resulting functions in Equation 4.31, give

$$E_i = \rho\left(\frac{x}{t}\right)_i^2 \qquad (4.33)$$

where the subscript i on the right-hand side indicates that the strain velocity x/t is that corresponding to the particular strain ϵ_i.

By means of Equation 4.33, values of the modulus E_i can be calculated for various selected velocities $(x/t)_i$. These velocities would be in the range covered by graphs such as Figure 4.11. The E_i's can then be plotted against the ϵ_i's corresponding to the same $(x/t)_i$'s for each striking velocity, as given by graphs of the type of Figure 4.11. Since $E = \partial\sigma/\partial\epsilon$, graphical integration

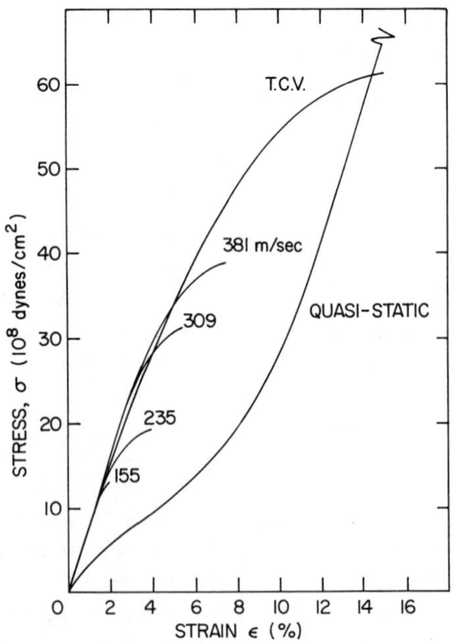

Fig. 4.12. Stress-strain curves of 210/5 nylon 66 yarn, in quasi-static test and for various transverse impact velocities. [After Reference 78.]

of the resulting curves gives σ as a function of ϵ, the stress-strain curve for each striking velocity. The family of such curves obtained by Petterson and Stewart for the nylon yarn, employing this technique, is shown in Figure 4.12.   An experimental quasi-static stress-strain curve for the same sample, obtained at 40% extension/min, is shown for comparison.

4.2 Fabric Systems

4.2.1 Introduction. An extension to two-dimensional structures, such as films and fabrics, of the theory for yarns given in the foregoing sections, has not thus far appeared in the published literature.   A theoretical treatment of the impact behavior of fabric systems, based on energy considerations, without regard to the mechanism of strain propagation, has, however, been developed by Mrs. Rogers.[84]   It will be of interest to review this treatment, not only as an analysis of the energetics involved when a fabric system is impacted normally (or transversely) by a high-speed missile, but also as an example of a different theoretical approach.

4.2.2 Experimental Background. As has been outlined in Chapter 3, the standard method[111] for the ballistic evaluation of fabrics for personnel armor consists essentially of impacting the fabric sample with a special, free-flying missile of measured velocity. The average of the striking velocities corresponding to five complete and five partial penetrations of the test panel, determined according to the procedure described in Chapter 3, is taken as the V_{50} ballistic limit. It is interpreted as the striking velocity (for the particular type of missile) at which, in a large number of impacts on specimens of the fabric sample, 50% will result in complete penetrations. A little reflection will disclose that the V_{50} velocity represents the limiting value at which the missile will barely puncture the fabric panel and witness plate, and come to rest in the latter. The kinetic energy of the missile corresponding to V_{50} is a measure of all of the energy which the panel-and-plate system is capable of absorbing, at the particular missile velocity.

It is intuitively evident that as fabrics of a particular construction and fiber type are made heavier, or as the number of layers of a particular fabric in a panel is increased, the V_{50} ballistic limit will be increased. In Figure 4.13 such a dependence of V_{50} on the areal or surface density of a panel of standard nylon armor fabric is shown, the variables being expressed in arbitrary units.[84]   Here the increase in areal density is achieved simply by increasing the number of sheets of fabric, one sheet having a value of 0.015 unit on the scale used here.

The parabolic character of the curve in Figure 4.13 suggests that, approximately,

$$V_{50} \propto \sqrt{\rho h} \qquad (4.34)$$

where ρh is areal density, ρ being the mass per unit volume (density) of the textile material, and h the effective thickness of the fabric panel. For two layers (each of one or more sheets) having ballistic limits $(V_{50})_a$ and $(V_{50})_b$, Equation 4.34 suggests the following relationship for the limit of the combination

$$(V_{50})_{ab} = \sqrt{(V_{50})_a^2 + (V_{50})_b^2} \qquad (4.35)$$

This equation is reported to have been found in experiment to hold approximately, at higher velocities, for many materials impacted with nondeforming missiles of the fragment-simulating type.[84]

From Equation 4.34, one would expect that for a panel consisting of n like sheets, the V_{50} limit would be given by $\sqrt{n}(V_{50})_1$, where $(V_{50})_1$ is the ballistic limit of one sheet.

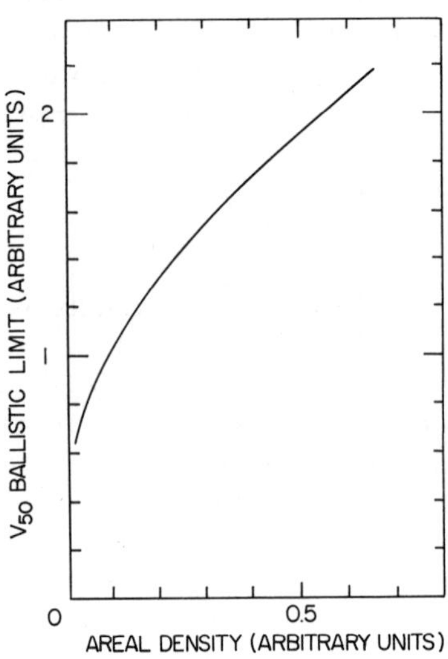

Fig. 4.13. Ballistic limit (V_{50}) of nylon armor fabric, as a function of areal density (number of layers) of panel. [After Reference 84.]

It is found, however, that for the case of 12 sheets of nylon armor fabric, the limit is

$$(V_{50})_{12} = 1.85 \, (V_{50})_1 \qquad (4.36)$$

which is considerably less than $\sqrt{12} \, (V_{50})_1$. The implication is that the parabolic relationship represented by Equation 4.34 does not hold for such low striking velocities as $(V_{50})_1$. In the following sections are outlined the considerations of Mrs. Rogers leading to a theoretical explanation of Equations 4.35 and 4.36, based on a relationship having more general validity than Equation 4.34.

4.2.3 Theory of Ballistic Limits. Basic equation for the ballistic limit. From Newton's second law of motion we have, for the retarding force acting on a projectile at any instant, as it penetrates a fabric panel,

$$F = - \frac{dT}{dx} \qquad (4.37)$$

where T is the kinetic energy of the projectile, and x is the distance it has penetrated into the target. Rearranging the terms of this equation, and integrating over the total distance of penetration h, we have

$$\int_0^h dx = -\int_{T_0}^{T_e} \frac{1}{F} \, dT$$

where T_0 is the kinetic energy with which the projectile strikes the target fabric, corresponding to $x = 0$, and T_e is the energy with which the projectile must emerge from the fabric in order barely to puncture the witness plate. Then

$$h = -\int_{T_0}^{T_e} \frac{1}{F} \, dT \tag{4.38}$$

where F is considered to be a function of T.

From what has been said earlier regarding the significance of V_{50}, T_0, and T_e, it is apparent that

$$T_0 = \tfrac{1}{2} M V_{50}^{2} \tag{4.39}$$

where M is the mass of the projectile. Hence Equation 4.38 takes the form

$$h = -\int_{\frac{1}{2}MV_{50}^{2}}^{T_e} \frac{1}{F} \, dT = \int_0^{\frac{1}{2}MV_{50}^{2}} \frac{1}{F} \, dT - \int_0^{T_e} \frac{1}{F} \, dT \tag{4.40}$$

The term $\int_0^{T_e} dT/F$ is a constant, so that if, after multiplication through by ρ, Equation 4.40 is differentiated with respect to V_{50}, one obtains

$$\frac{d(\rho h)}{dV_{50}} = \rho \frac{1}{F_0} M V_{50}^{2} \tag{4.41}$$

where, now, F_0 represents the retarding force acting at the infinitesimally thin front surface of the panel.[84]   From this equation we have

$$F_0 = M\rho V_{50} \frac{dV_{50}}{d(\rho h)} \tag{4.42}$$

For the higher areal densities ρh (and striking velocities) the data of Figure 4.13 can be represented by the function

$$V_{50}^{2} = 0.42 + 6.72\rho h; \quad \text{when } \rho h \geq 0.2 \tag{4.43}$$

From Equations 4.42 and 4.43, then

$$F_0 = 3.36 M\rho; \quad \text{when } \rho h \geq 0.2 \tag{4.44}$$

This relationship indicates that, at the higher values of V_{50} and ρh, the initial retarding force is independent of the striking velocity.

By means of graphical differentiation of the curve in Figure 4.13 and the application of Equation 4.42, values of $F/N\rho$ (which is proportional to the retarding force for a given missile) can be calculated. The results show that at the lower striking velocities for penetration V_{50}, the retarding force becomes greater.[84] To fit the data at the lower velocities, without introducing serious error at the higher ones, Equation 4.44 is modified to the form

$$F_0 = M\rho p \left(1 + \frac{1}{qV_{50}^{\,2}}\right) \tag{4.45}$$

where $p \approx 3.3$ and $q \approx 4.5$. Mrs. Rogers makes the implicit assumption that Equation 4.45 not only holds at the front surface of the target, but represents the relationship between the instantaneous retarding force F and the missile velocity v during penetration.* Thus

$$F = -\frac{dT}{dx} = M\rho p \left(1 + \frac{1}{qv^2}\right) \tag{4.46}$$

Since $T = Mv^2/2$, $dT = Mv\,dv$, so that one can rearrange Equation 4.46 as follows:

$$\frac{Mv\,dv}{1 + \dfrac{1}{qv^2}} = -M\rho p\,dx \tag{4.47}$$

These expressions can be readily integrated, yielding

$$\frac{1}{q} + v^2 - \frac{1}{q}\ln\left(\frac{1}{q} + v^2\right) = -2\rho p x + \text{const} \tag{4.48}$$

Evaluation between the limits $v = V_{50}$ at $x = 0$, and $v = v_e$ at $x = h$, results finally in the relationship

$$V_{50}^{\,2} - \frac{1}{q}\ln\left(1 + qV_{50}^{\,2}\right) = 2\rho p h + v_e^2 - \frac{1}{q}\ln\left(1 + qv_e^2\right) \tag{4.49}$$

where v_e is the velocity with which the fragment-simulating missile must emerge from the fabric panel in order barely to puncture the witness plate.

Equation 4.49 represents an expression by means of which the ballistic limit V_{50}, of a fabric panel can, in principle, be calculated from the experimental parameters ρh and v_e, and the curve-fitting constants p and q. Equation 4.49 is to be considered a

---

* For the penetration process, p and q may be expected to have effective numerical values different from those quoted above.

refined version, based in part on dynamic theory, of the purely empirical Equation 4.43 for the curve of Figure 4.13. The expression $v_e^2 - (1/q) \ln (1 + qv_e^2)$ in Equation 4.49 is a constant for a particular type of missile. It and the "corrective" term on V_{50}^2, $(1/q) \ln (1 + qV_{50}^2)$, taken together (though not constant), may be thought of as corresponding to the constant 0.42 in Equation 4.43. Equation 4.35 evidently can be derived from the approximate Equation 4.43, if the constant 0.42 is neglected.

Ballistic limit of two layers at high velocity. The theoretical Equation 4.49 will now be applied to the derivation of Equation 3.35, considered as an empirical expression for the V_{50} limit of two layers, each of known ballistic limit. A layer is taken to consist of one or more sheets of fabric, considered as a unit. Let the areal densities of the two layers be $(\rho h)_a$ and $(\rho h)_b$, so that the areal density of the combination is $(\rho h)_a + (\rho h)_b$.* The expression for the combination then becomes, from Equation 4.49,

$$(V_{50})_{ab}^2 - \frac{1}{q} \ln \left[1 + q(V_{50})_{ab}^2 \right] = 2p \left[(\rho h)_a + (\rho h)_b \right] + v_e^2 - \frac{1}{q} \ln (1 + qv_e^2)$$

$$(4.50)$$

But Equation 4.49 also gives expressions involving the ballistic limits $(V_{50})_a$ and $(V_{50})_b$ for each of the two layers separately:

$$(V_{50})_a^2 - \frac{1}{q} \ln \left[1 + q(V_{50})_a^2 \right] = 2p(\rho h)_a + v_e^2 - \frac{1}{q} \ln (1 + qv_e^2)$$

$$(4.51)$$

and

$$(V_{50})_b^2 - \frac{1}{q} \ln \left[1 + q(V_{50})_b^2 \right] = 2p(\rho h)_b + v_e^2 - \frac{1}{q} \ln (1 + qv_e^2)$$

$$(4.52)$$

Adding Equations 4.51 and 4.52, subtracting the result from Equation 4.50, and rearranging terms lead to

$$(V_{50})_{ab}^2 - \frac{1}{q} \ln \left[1 + q(V_{50})_{ab}^2 \right]$$

$$= (V_{50})_a^2 + (V_{50})_b^2 - v_e^2 + \frac{1}{q} \ln \left[\frac{1 + qv_e^2}{\left[1 + q(V_{50})_1^2 \right] \left[1 + q(V_{50})_2^2 \right]} \right]$$

$$(4.53)$$

---

* The present derivation is a generalization of that of Mrs. Rogers,[84] who considered both layers to have the same ρ, that is, to be composed of the same material (fiber type). This restriction is unnecessary.

For relatively high V_{50} velocities, in the range 1 to 2, in the arbitrary units used here, v_e (≈ 0.35 unit) and certainly, its square, are negligible. Thus we may write to a good approximation

$$(V_{50})_{ab}^2 = (V_{50})_a^2 + (V_{50})_b^2 + R \qquad (4.54)$$

where

$$R = \frac{1}{q} \ln \left[\frac{1 + q(V_{50})_{ab}^2}{\left[1 + q(V_{50})_a^2\right]\left[1 + q(V_{50})_b^2\right]} \right]$$

The argument of the logarithm here has a fractional value for all values of $(V_{50})_a$ and $(V_{50})_b$. Thus, the logarithmic term R represents the negative residual between the $(V_{50})_{ab}^2$ given by Equation 4.54, and the approximate empirical sum $(V_{50})_a^2 + (V_{50})_b^2$. For the range of velocities considered here, however, R is sufficiently small* that it may be neglected, without introducing a discrepancy larger than the uncertainty in the experimental data themselves. Thus, we obtain

$$(V_{50})_{ab} = \sqrt{(V_{50})_a^2 + (V_{50})_b^2} \qquad (4.55)$$

in agreement with Equation 4.35.

An idea of the extent to which neglecting R alters calculated values is given by the following example: For $(V_{50})_a = (V_{50})_b = 1.225$ units, and $q = 3$, the logarithm defining R has the value -1.2, so that $R = -0.4$. From Equation 4.54, then, taking R into consideration, we obtain $(V_{50})_{ab} = 1.61$ units. Neglecting R (calculating $(V_{50})_{ab}$ from Equation 4.55), we obtain 1.73 units. Actually, the result from Equation 4.55 agrees as well with the empirical value indicated by Figure 4.13, about 1.66 units, as does the result from Equation 4.54.

Ballistic limit of twelve sheets. Equation 4.49 can be, likewise, applied directly to the derivation of an expression for the ballistic limit of n like sheets, in terms of the limit of a single sheet. For such a single sheet, having areal density ρh, and the ballistic limit $(V_{50})_1$,

$$(V_{50})_1^2 - \frac{1}{q} \ln \left[1 + q(V_{50})_1^2\right] = 2\rho\rho h + v_e^2 - \frac{1}{q} \ln (1 + qv_e^2)$$

$$(4.56)$$

---

* The rationale adopted here differs from that of Mrs. Rogers, who states that the whole logarithmic term R in Equation 4.54 approaches zero as $(V_{50})_a$ and $(V_{50})_b$ increase to very large values. This statement is not valid, since, under these conditions, the argument $\to 0$, and $R \to -\infty$.

Similarly, Equation 4.49 gives for n of the same sheets, having the limit $(V_{50})_n$,

$$(V_{50})_n^2 - \frac{1}{q} \ln\left[1 + q(V_{50})_n^2\right] = 2n\rho p h + v_e^2 - \frac{1}{q} \ln(1 + qv_e^2)$$

(4.57)

Multiplying Equation 4.56 by n, we obtain from the two equations

$$(V_{50})_n^2 - \frac{1}{q} \ln\left[1 + q(V_{50})_n^2\right]$$

$$= n(V_{50})_1^2 - \frac{n}{q} \ln\left[1 + q(V_{50})_1^2\right] - (n-1)v_e^2 + \frac{(n-1)}{q} \ln(1 + qv_e^2)$$

(4.58)

Assuming that $(V_{50})_1$ is sufficiently large that the terms in v_e can be ignored, we have from Equation 4.58

$$(V_{50})_n^2 - \frac{1}{q} \ln\left[1 + q(V_{50})_n^2\right] = n(V_{50})_1^2 - \frac{n}{q} \ln\left[1 + q(V_{50})_1^2\right]$$

(4.59)

The logarithmic terms in this equation may be represented by the series defined by the formula

$$\ln(1 + x) = x - \frac{x^2}{2} + \frac{x^3}{3} - \cdots$$

For small values of x the logarithm is approximated by the first two terms of the series. When this substitution is made, Equation 4.59 becomes

$$(V_{50})_n^2 - \frac{1}{q}\left[q(V_{50})_n^2 - \frac{q^2(V_{50})_n^4}{2}\right] = n(V_{50})_1^2 - \frac{n}{q}\left[q(V_{50})_1^2 - \frac{q^2(V_{50})_1^4}{2}\right]$$

(4.60)

or,

$$\frac{q(V_{50})_n^4}{2} = n\frac{q(V_{50})_1^4}{2}$$

Taking the fourth root on both sides of this equation yields

$$(V_{50})_n = \sqrt[4]{n}(V_{50})_1$$

(4.61)

For the case of n = 12,

$$(V_{50})_{12} = \sqrt[4]{12}(V_{50})_1 = 1.86(V_{50})_1$$

(4.62)

This relationship can be seen to be in very good agreement with the empirical Equation 4.36.

The precision of the foregoing derivations evidently depends on the validity with which, in the particular application, the assumption $v_e = 0$ can be made, and the other indicated approximations, in connection with Equations 4.54 and 4.59, adopted. Regardless of this point, the derivations provide an insight into the physical factors that come into play in the penetration of a fabric panel by a ballistic missile. The agreement of Equations 4.55 and 4.62 with empirical findings indicates that the postulated physical relationships are substantially correct.

Chapter 5

RELATED DYNAMIC PROPERTIES

5.1 Introduction

The impact properties of a material may be considered to be special aspects of its dynamic mechanical properties, in the general sense. More particularly, however, the term dynamic properties has come to refer to the behavior of a material under very rapid cyclic stressing or straining at amplitudes sufficiently small so that rupture does not occur, and in most cases, so that not even much permanent deformation results. Impact properties, on the other hand, have been taken to be those related to the rapid permanent deformation, and usually rupture, of a material (as discussed in the preceding chapters). The same molecular mechanisms and parameters that determine the dynamic properties of a material would undoutedly also define its behavior during an impact process, at least in the initial phase. It is, therefore, of interest to consider briefly, in connection with impact phenomena in textiles, the nature of the elementary dynamic properties.

It was pointed out in Chapter 2 that in polymeric materials such as textiles, elastic and viscous effects occur together. If, then, a force is applied to a textile filament or yarn, tending to stretch it, this force must overcome combined elastic and viscous reactions. According to Equations 2.1, 2.2, and 2.3, the elastic resistance is

$$a\sigma = aE\left(\frac{\ell - \ell_o}{\ell_o}\right) = \frac{a}{\ell_o} Es \tag{5.1}$$

where $s = \ell - \ell_o$ is the displacement (in units of length) of the end of the sample, and the other symbols have the meanings given in Chapter 2.

If we assume that the viscous process conforms to the law for an ideal fluid, that is, that stress is proportional to the rate of deformation or extension, this part of the resistance will be given by the expression

$$a\mu \frac{d}{dt}\left(\frac{\ell - \ell_o}{\ell_o}\right) = \frac{a}{\ell_o}\mu \frac{ds}{dt} \tag{5.2}$$

where μ is a proportionality constant characterizing the internal friction associated with the viscous behavior of the material.

The total resistance to extension of a textile specimen is given approximately, then, by the equation of motion:

$$\frac{a}{\ell_o}\mu\,\frac{ds}{dt} + \frac{a}{\ell_o}\,Es \;=\; F(t) \tag{5.3}$$

where $F(t)$ is the applied force, in equilibrium with the resisting forces, at every instant. The behavior of a viscoelastic body represented by Equation 5.3 is the same as that of a spring connected in parallel with a dashpot, that is, a leaky piston in a cylinder. Such a mechanical model, the Voigt, as it is called, is shown in Figure 5.1.

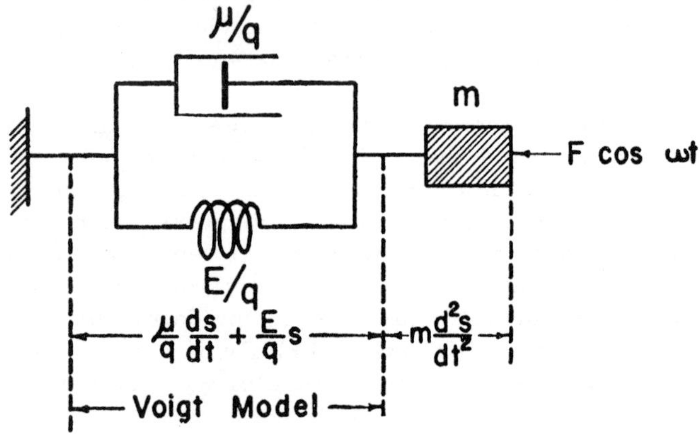

Fig. 5.1. Idealized mechanical model of vibrometer system, with Voigt model representing viscoelastic yarn.[49]

5.2 Measurement of Dynamic Properties

The parameters E and μ, when measured under rapid cyclic loading, and others derived from them, specifically represent or constitute the dynamic properties of a material. When measured in this manner, E is called the dynamic Young's (or stretch) modulus. The quantity μ has been called the coefficient of internal friction, among other designations.

Methods thus far developed for the measurement of the dynamic properties of textiles have generally involved the use of a vibrating mass to which the specimen is attached.[49,50,108]*   A schematic diagram of typical apparatus is shown in Figure 5.2. The principle and operation of such a stretch-vibrometer have been described by Lyons[50] as follows: ". . . the method consists of passing a small oscillating current of known frequency through

---

*Reference 108 cites literature describing various versions of this apparatus developed during the 1950's.

a coil which rests in the field of a strong electromagnet and is
supported coaxially by two equal lengths of the test string, which
are necessarily under tension. The passage of the current through
the coil causes it to vibrate axially, thus alternately elongating
and relaxing the specimen string. The amplitude of vibration is
measured by means of a micrometer microscope focused on a
mark on the vibrating system. The stretch-vibrometer has been
so calibrated that for a given length of sample (usually 25.4 cm.)
readings of the vibrating-coil current, its frequency, and the amplitude of vibration provide the necessary data from which the desired dynamic parameters can be calculated. Measurements are
made at resonance. Typical extremes in amplitude of vibration
are 0.080 cm. (0.3 percent) for a nylon cord, and 0.010 cm. (0.04
percent) for glass cord."

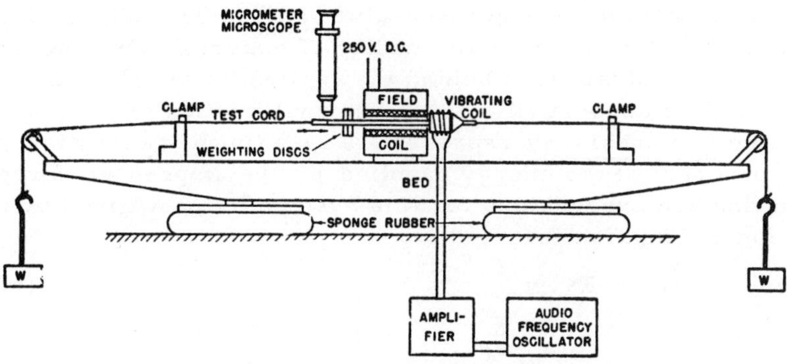

Fig. 5.2. Schematic diagram of stretch-vibrometer and
auxiliary equipment.[50]

Because the rapidly accelerating and decelerating coil-and-
weight assembly, as well as the two lengths of specimen, form
part of the vibrating system, the mass of this assembly must be
taken into consideration for the complete equation of motion. The
reaction of this mass, considered as another force resisting the
applied one, and being given by the product of mass and acceleration, must be added to the resistance presented by the specimen,
as given by Equation 5.3. Furthermore, as the force F(t) applied
by the vibrating coil is not a steady pull, but one varying sinus-
oidally (or very nearly so) with time, an explicit expression
can be introduced for it. Thus, we have for the equation of
motion underlying this method:

$$m \frac{d^2 s}{dt^2} + \frac{\mu}{q} \frac{ds}{dt} + \frac{Es}{q} = F \cos \omega t \qquad (5.4)$$

where m is the effective mass of the vibrating assembly; q is
$\ell_0/2a$, ℓ_0 being the initial length of each of the matched halves of

the specimen; and F is the amplitude of the applied force, which
varies with a frequency of $\omega/2\pi$ cycles/sec. This equation reflects the simplifying observation that instrumental, frictional
losses are negligible. The force amplitude F is obtained from
a calibration of the instrument giving force as a function of the
current through the field coil.

The solution of Equation 5.4 leads to the following formulas
for the basic dynamic properties:

$$E = qm\omega_0^2 \tag{5.5}$$

$$\mu = \frac{qF}{s_m\omega_0} \tag{5.6}$$

The quantity ω_0 represents the value of ω at resonance, while s_m
is the maximum amplitude (at resonance).

It was pointed out in the discussion of viscoelasticity in Chapter 2, that in the rapid cyclic loading of materials showing such
behavior, the strain lags behind the applied force. This phenomenon, which is associated with the internal friction of the material,
is called hysteresis. It represents a loss (in the generation of
heat) of a part of the energy supplied by the impressed force.
Following are expressions for this hysteretic loss, given by the
solution of Equation 5.4:

$$H_c = Fs_m\pi \tag{5.7}$$

$$H_{c,v} = \frac{\mu s_m^2\omega_0\pi}{\ell_o^2} \tag{5.8}$$

$$= \frac{Fs_m\pi}{2a\ell_o} \tag{5.9}$$

where H_c is the total energy loss per cycle, and $H_{c,v}$ is the hysteretic constant (energy loss per cycle per unit volume).

5.3 Energy-Loss Functions

5.3.1 Energy Loss at Unit Strain-Amplitude. The hysteretic
constant $H_{c,v}$ evidently applies to specimens of a particular
length ℓ_o strained at a particular displacement s_m, since both
of these variables appear in Equations 5.8 and 5.9, and no other
quantities in the expressions compensate for them. It is desirable, however, to have a parameter that is characteristic of the
material, not explicitly involving the experimental conditions
under which it is measured. To this end, a quantity that does
not depend on the strain amplitude s_m/ℓ_o, and that is hence more
nearly a material constant than $H_{c,v}$, was introduced.[49]   This
quantity has been called, preferably, the energy loss at unit strain-

amplitude. Since, as Equation 5.8 shows, energy loss is proportional to the square of the strain-amplitude $(s_m/\ell_o)^2$, this energy-loss function is obtained by dividing the hysteretic constant by $(s_m/\ell_o)^2$. Energy loss at unit strain-amplitude is thus given by the expression $\mu\omega_0\pi$. When energy losses are thus adjusted (in effect) to a uniform strain-amplitude, in this case unity, comparisons can be made between materials tested at different strain-amplitudes. Ideally, values obtained for different samples bear the same relations to each other as if they had been tested at the same amplitude. Lyons cautions, however: "Because it is found experimentally that even energy losses referred to unit strain-amplitude are not, in general, strictly independent of strain-amplitude itself, the amplitude range should not be too wide."

5.3.2 Energy Loss at Unit Stress-Amplitude. In most mechanical applications of textiles, the textiles are the load-carrying elements. Different materials in a particular service in general might experience widely varying strains, because of differences in properties. This results when the forces imposed on an element of one material are the same as those imposed on an element of a replacement material, and no compensation is made by altering the amount of material used. In applications involving cyclic tension, such replacements would generally introduce different strain-amplitudes, but stress-amplitudes would tend to be the same. If, for example, one textile material in a pneumatic tire is replaced by another, approximately cord for cord, the individuals of the two samples could be expected to experience cycles of nearly equal stress-amplitude, under similar operating conditions. Where such are the applications, it is unrealistic to make comparisons on the basis of unit strain-amplitude; rather, they should be made in terms of energy loss at unit stress-amplitude.

The total of the stresses in a body at any instant arise from viscous as well as elastic effects. Effectively, the total stress-amplitude is given by the so-called "vector" sum of the elastic component, Es_m/ℓ_o, and the viscous, $\mu\omega_0 s_m/\ell_o$, that is, by $\sqrt{(Es_m/\ell_o)^2 + (\mu\omega_0 s_m/\ell_o)^2}$. As in the case of energy loss at unit strain-amplitude, dividing the hysteretic constant (as given in Equation 5.8) by the square of the total stress-amplitude yields the energy loss at unit stress-amplitude. Thus, we obtain for this function, $\mu\omega_0\pi/(E^2 + \mu^2\omega_0^2)$. In substituting actual values of E, μ, and ω_0 in this formula, it is found that for textile materials, E^2 is much larger than $\mu^2\omega_0^2$, usually by a factor of 1000 or more. The precision with which E^2 can be calculated does not warrant retaining the relatively much smaller addend $\mu^2\omega_0^2$. Accordingly, energy loss at unit stress-amplitude has been calculated[49] as $\mu\omega_0\pi/E^2$. This quantity, like that referred to unit strain-amplitude, is not a perfect constant, but varies when strain-amplitudes cover a wide range.

5.3.3 Loss Tangent.   On the right-hand side of Equation 5.4 time is represented by an angle, one of ω radians corresponding to 1 sec; $2\pi/\omega$ sec is the duration of 1 cycle. The interval of time by which, as noted above, strain lags behind applied force because of hysteresis also can be represented by an angle θ, called the loss or phase angle. This is to say, in a cycle the strain reaches its maximum value θ/ω sec after the impressed force reaches its maximum and starts to decline.   This lag is on the order of a small fraction of a second in the type of measurements considered here.

The tangent of the loss angle, or loss tangent, tan θ, is another parameter that has been used to characterize the hysteresis of a material. The expression for tan θ, the derivation of which is beyond the scope of the present treatment, is $\mu\omega/E$. At resonance, tan $\theta = \mu\omega_0/E$. The loss tangent is not an absolute measure of the hysteretic energy, but is proportional to the ratio of the energy dissipated (represented by $\mu\omega$) to the maximum stored energy (represented by E) in a cycle. It can be seen that for a series of materials having equal moduli E, the loss tangents will increase with increasing μ, provided the dynamic-property measurements are made at the same angular frequency ω. That is, other parameters being uniform, a high loss tangent reflects high internal friction.

5.4 Characteristics of Dynamic Stretch Moduli

5.4.1 Absence of Frequency-Dependence.   Lyons and Prettyman[50] conducted an experiment on cotton tirecord in which 19 different combinations of specimen length ℓ_0 and vibrating mass m were used with a constant static tension of 2.0 kg on the specimen. They found values of the modulus E, calculated by means of Equation 5.5, varying from 9.6 to 10.7×10^{10} dynes/cm^2. The dispersion of values appeared to be essentially random; it could not be correlated with the variations in ℓ_0 or m, nor with those in the resonant frequency $\omega_0/2\pi$ that resulted when ℓ_0 or m was changed. These results confirmed the expectation that E is a parameter of the material in a particular state, not dependent on the instrumental factors ℓ_0 and m used in its measurement. Furthermore, the experiment demonstrated that, over the range of frequencies employed (70 to 300 cycles/sec), E is independent of frequency. A very similar experiment employing a high-tenacity nylon 66 monofil led to the same conclusion.

This lack of strong dependence of the dynamic modulus on frequency or strain rate was also found by Tipton[108] in measurements on viscose rayon yarns in wide ranges of frequency and

strain-amplitude. Representative results are shown in the upper part of Figure 5.3.* It can be seen that the trends with frequency (if any) in modulus values found at any given strain-
amplitude, are not very consistent among the various strain-
amplitudes. The dispersion in modulus values (≈ 1 dyne/cm^2)
is about the same as that found by Lyons and Prettyman.

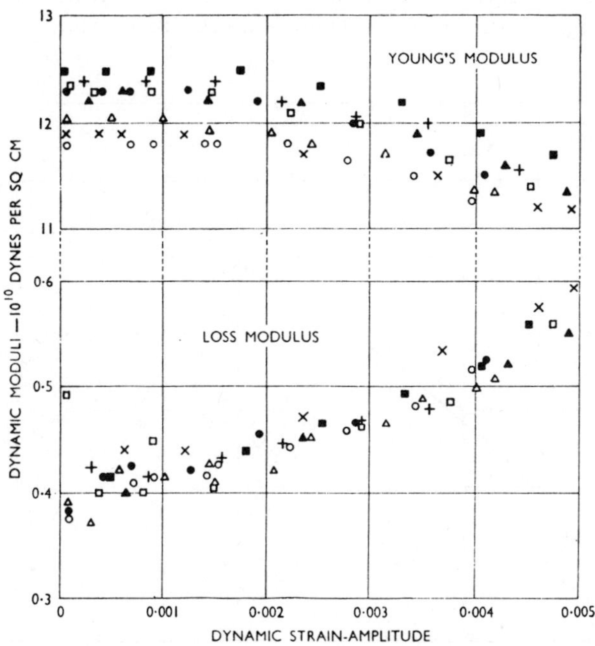

a Continuous-filament viscose rayon single yarn.

○	18 cycles per sec	●	22 cycles per sec
△	41 cycles per sec	□	80 cycles per sec
▲	56 cycles per sec	■	108 cycles per sec
+	115 cycles per sec	×	150 cycles per sec

Fig. 5.3. Dynamic Young's modulus and loss modulus of
viscose yarn as functions of strain amplitude, at various
frequencies.[108]

5.4.2 Dependence on Static Tension or Mean Strain.   When
measurements were made on the cotton cord, discussed in the
foregoing paragraphs, with a constant length, ℓ_0 and mass m,
but with different static tensions, a substantial, systematic increase in the resonant frequency was found, as the tension was
increased. Thus, the modulus as given by Equation 5.5 showed
a corresponding increase with tension. The results are presented

____

* For the significance of the loss modulus, for which data are
plotted in Figure 5.3, see Section 5.5.1.

graphically in Figure 5.4. The increase in the dynamic modulus can be considered to reflect the change in slope of the quasi-
static force-extension curve of the cotton cord.   Such a curve
is concave-upward, so that the slope at each point (corresponding to what has been called the "tangent modulus") increases
with tension. In the dynamic experiments, then, as the tension
is increased, the specimen may be considered as being rapidly
extended and compressed around equilibrium positions associated with higher and higher stress-to-strain ratios.

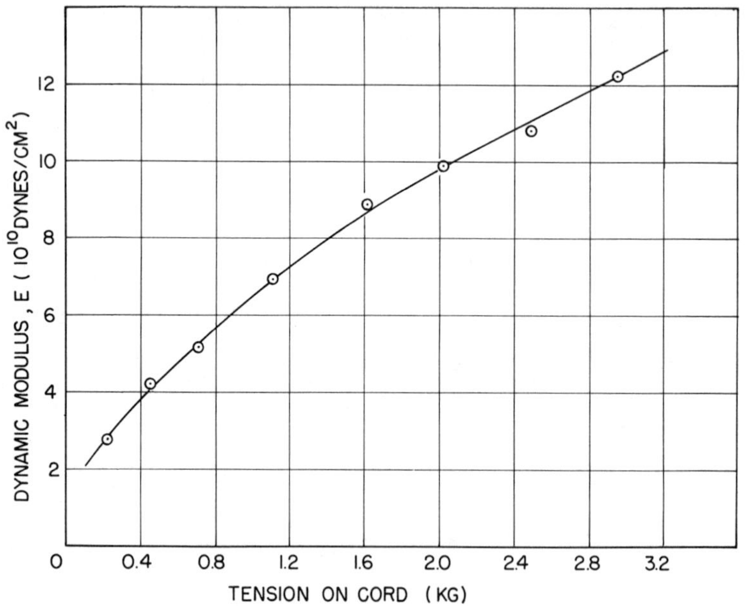

Fig. 5.4. Effect of initial static tension in specimen on the
dynamic modulus of 11/4/2 cotton cord. [After Reference 50.]

Tipton found similar increases in the dynamic moduli of yarns
of a wide variety of fiber types when the moduli were plotted as
functions of mean strain, that is, the strain that is imposed on
the specimen when it is mounted in the apparatus preparatory to
testing. His results are shown in Figure 5.5. The cellulose acetate, viscose, nylon, and Terylene samples were low-twist continuous-filament yarns. Their behavior, as shown by the curves
in Figure 5.5, can be considered to be substantially that of the
filamentous materials themselves. The curves for the natural
staple-fiber yarns, wool, cotton, and linen (like that of the cotton cord in Figure 5.4) reflect the influence of yarn structure, as
well as fiber properties. Since static tension is directly, if not
linearly, related to mean strain, Tipton's results may be con-

sidered as substantially confirmatory extensions of those based on varying static tension.

Tipton makes the following observation concerning this behavior: "There is generally an increase in modulus as the fibers are strained. This is a result of an increase in orientation of the fiber material with strain, which also shows up as an increase in the optical anisotropy. . . . The more highly oriented fibers, whether natural or man-made, show the highest modulus values, and also the highest rate of increase of modulus with mean strain."

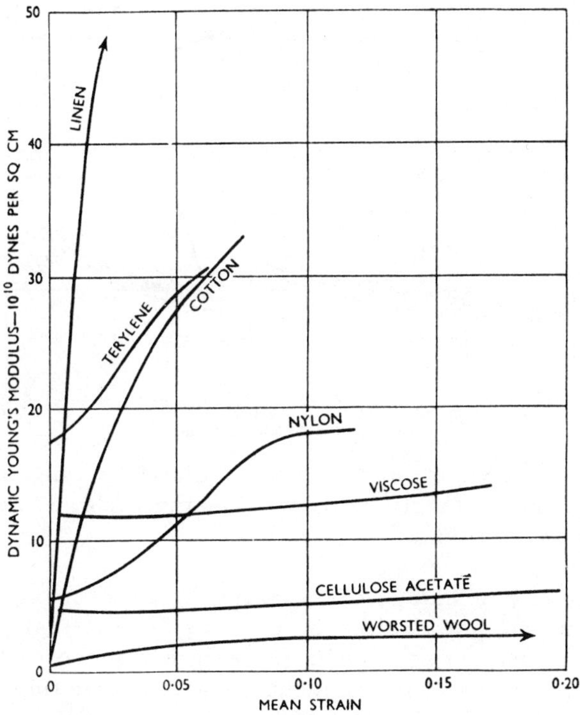

Fig. 5.5. Effect of initial static (mean) strain on the dynamic moduli of various textile materials.[108]

5.4.3 Influence of Strain-Amplitude. When dynamic measurements are made on yarns at various strain-amplitudes, the moduli obtained are found to fall off slightly with increasing amplitude. This is shown by Tipton's data for a continuous-filament, single yarn in Figure 5.3. He also found this behavior in nylon, Terylene, and cellulose acetate samples. In the upper part of Figure 5.6 are shown his results on the moduli of a low-twist single and a cabled continuous-filament viscose yarn at various mean strains from 0.01 to 0.15. The positions of the curves for each mean strain reflect the trend shown for the viscose sample in Figure 5.5.

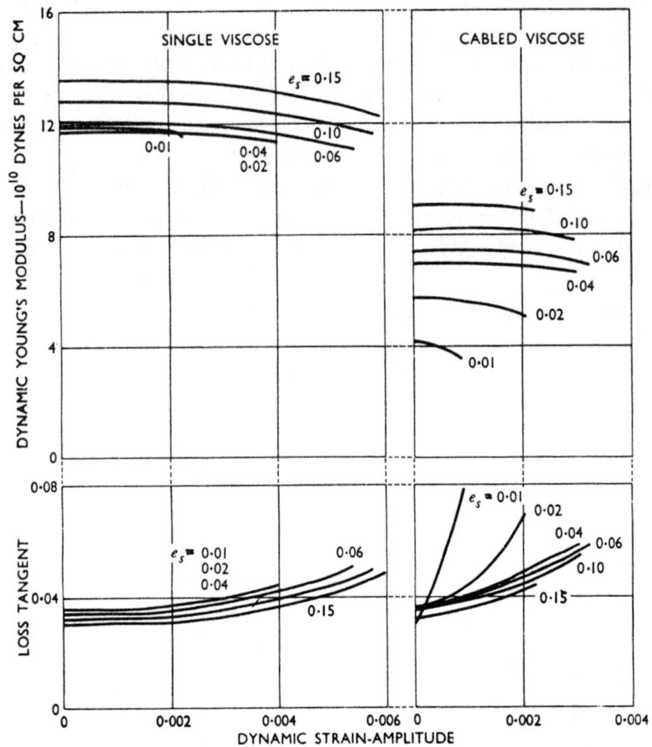

Fig. 5.6. Dynamic Young's modulus and loss tangent of viscose yarns as functions of strain-amplitude, at various static (mean) strains.[108]

On the basis of percentage of the value of each curve at zero strain-amplitude, the decline in modulus is practically the same for all mean strains, at least in the single viscose yarn.

The decline in the dynamic modulus with increasing strain-amplitude Tipton has ascribed to interfiber friction effects. At the lower strain-amplitudes the segments of adjacent fibers in contact tend to stick together. As the strain amplitude is increased, and with it, the dynamic stress, the static frictional force is overcome, and fiber segments slip past one another. Thus, there result relatively larger over-all extensions of the yarn for a given applied force; effectively, the dynamic modulus is reduced. This effect would be expected to be most pronounced in twisted and cabled yarns, especially those composed of staple fibers, where the extensibility of the assembly is controlled in large degree by interfiber friction. As the mean strain in a twisted yarn is increased, it becomes laterally more compact; that is, adjacent fibers come into more intimate contact. The normal interfiber forces, it is postulated, are increased so that larger applied axial forces are required to obtain a given extension. This represents at a particular strain-amplitude, higher moduli as the mean strain is increased, an effect that can be seen in Figure 5.6.

5.4.4 Dynamic and Quasi-Static Moduli Compared. In Table 5.1 are given the dynamic stretch moduli of 15 samples of textile materials, glass fiber, and steel wire, in the order of descending values of the moduli, as reported by Lyons.[49] Each of the dynamic parameters E is an average of at least three measurements made with different vibrating masses and specimen lengths. A uniform tensile load of 2.0 kg was applied to all the specimens except the steel cord, on which a 2.9-kg load was used. In column 7 is given the resultant static tension in each sample during the dynamic experiment.

Loading and unloading curves for the 15 samples, for at least two complete cycles, were also obtained in conventional low-speed quasi-static tests. The number of strands in each specimen were selected in each case so as to duplicate the cross-sectional area of the sample in the dynamic test. From the mean slope of the curves in the second cycle or loop for each sample, a parameter taken to be the "static modulus" was calculated. This is entered in column 5 of Table 5.1.

While the "static modulus" derived in this manner is a rather arbitrary quantity, and not a true modulus, its comparison with the dynamic is of interest. In this regard, Lyons says: "It will be observed that for all samples except the steel monofil the ratio of dynamic to static modulus. . . is substantially greater than unity. This may be interpreted as indicating that during the slow static loading, quasi-elastic, if not plastic flow occurs, thus contributing to the observed elongation corresponding to a given load. The computed static modulus is thus smaller than it would be, were this contribution absent. In the dynamic test, there is insufficient time for the flow and orientation of the molecules of the filaments, or for the rearrangement and shift of the fibers and filaments of yarns and cords, before the applied force is reversed. The more plastic the material, the greater the ratio of dynamic to static modulus. Thus for the steel monofil, for which, in this static test, there was little opportunity for plastic flow, the ratio is a minimum, while for the newer synthetic organic filaments — Fiber A, Vinyon N, and Velon — which exhibit pronounced elongation in standard creep experiments, the maximum ratios are found. The role played by plastic flow in determining the ratios of dynamic to static moduli for various filament samples is eloquently shown by comparing the ratios for Vinyon N, nylon, and rayon yarns in Table 5.1 with the creep of these yarns." In creep tests on these three samples, the continuous-filament viscose, with a ratio of 1.86, was found to have

Table 5.1. Dynamic Moduli of Cords, Yarns, and Monofils Compared to Quasi-Static Moduli[49]

(Temperature Range: 21° - 25°C)

Description of Sample		Cross-sectional Area (cm^2)	Stretch Modulus (10^{10} dynes/cm^2)		Ratio of Dynamic to Static Modulus	Static Tension on Specimen (kg /mm^2)
Material	Construction		Dynamic E	Static (2nd loop)		
Steel	Monofil:6-mil	0.00018	213	203	1.05	112.8
Steel	6 x 3 x .0058-in cord	.0036	106	83	1.28	8.1
Fiberglas	7-ply cord	.0026	53.9	26.2	2.06	7.8
Ramie	3-ply cord	.0018	31.9	22.2	1.44	11.3
Rayon(viscose)	1100-den yarn (2 strands)	.0016	19.7	10.6	1.86	12.7
Fiber A(du Pont)*	300-den yarn (6 strands)	.0017	16.4	6.1	2.69	11.9
Rayon(viscose)	1100/2 cord	.0018	15.2	7.4	2.06	11.3
Fiber V(du Pont)**	75/16/2 cord	.0022	10.9	6.4	1.70	9.2
Vinyon N	300-den yarn; Type NORU (6 strands)	.0016	10.6	4.4	2.41	12.7
Cotton	11/4/2 cord	.0032	10.2	4.6	2.22	6.4
Nylon 66	210-den yarn (9 strands)	.0018	9.8	4.8	2.04	11.3
Nylon 66	210/3/3 cord	.0019	8.7	3.9	2.23	10.7
Nylon 66(high-tenacity)	Monofil (2 strands)	.0024	8.0	4.3	1.86	8.5
Velon	Monofil (3 strands)	.0021	6.4	2.3	2.78	9.7
Nylon 66	Monofil; 12-mil (3 strands)	.0020	6.0	2.6	2.31	10.1

*   Prototype of Orlon

** Prototype of Dacron

the least creep, while Vinyon N, having a ratio of 2.41, had the greatest creep; nylon, with a 2.04 ratio, was intermediate in creep.

5.5 Internal Friction and Energy Losses

5.5.1 Dependence of μ on Frequency. Values of the coefficient μ calculated according to Equation 5.6 from measurements made at different resonant frequencies show a decline as the frequency is increased. In Figure 5.7 are plotted such data for a nylon 66 monofil tested at more than 20 different combinations of specimen length and vibrating mass.[49] Similar results on a variety of

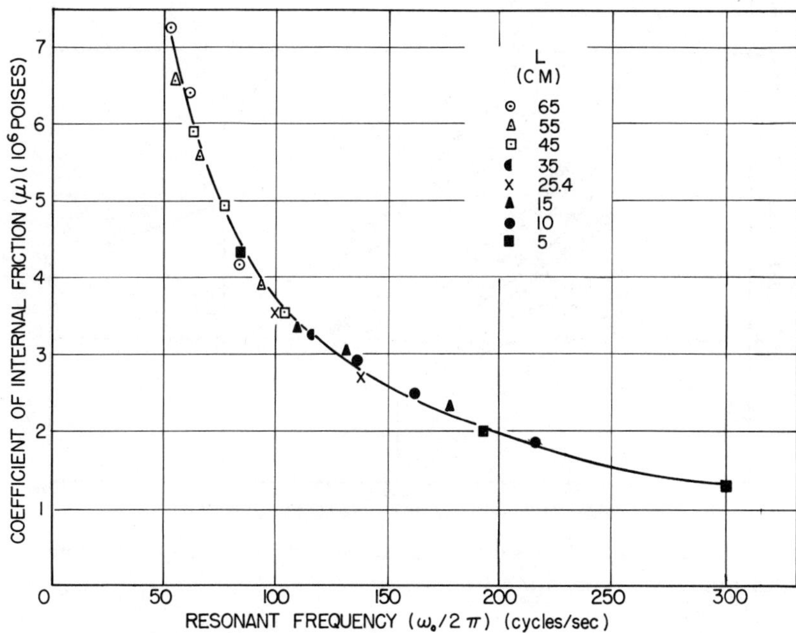

Fig. 5.7. Coefficient of internal friction of a high-tenacity nylon 66 monofil as a function of frequency, for various specimen lengths. [After Reference 49.]

fiber types, but with fewer individual data on each sample, are plotted in Figure 5.8. In most cases the experimental points conform very well to the hyperbolic law: $\mu\omega_0$ = constant. Experimentally, this law implies that as ℓ_o for a particular sample is changed (so as to alter q in Equation 5.6), F or s_m vary in such a way that the term qF/s_m remains substantially constant. The constant product $\mu\omega_0$ will be recognized as being the numerator of the fraction defining the loss tangent; $\mu\omega_0$ is sometimes called the loss modulus.

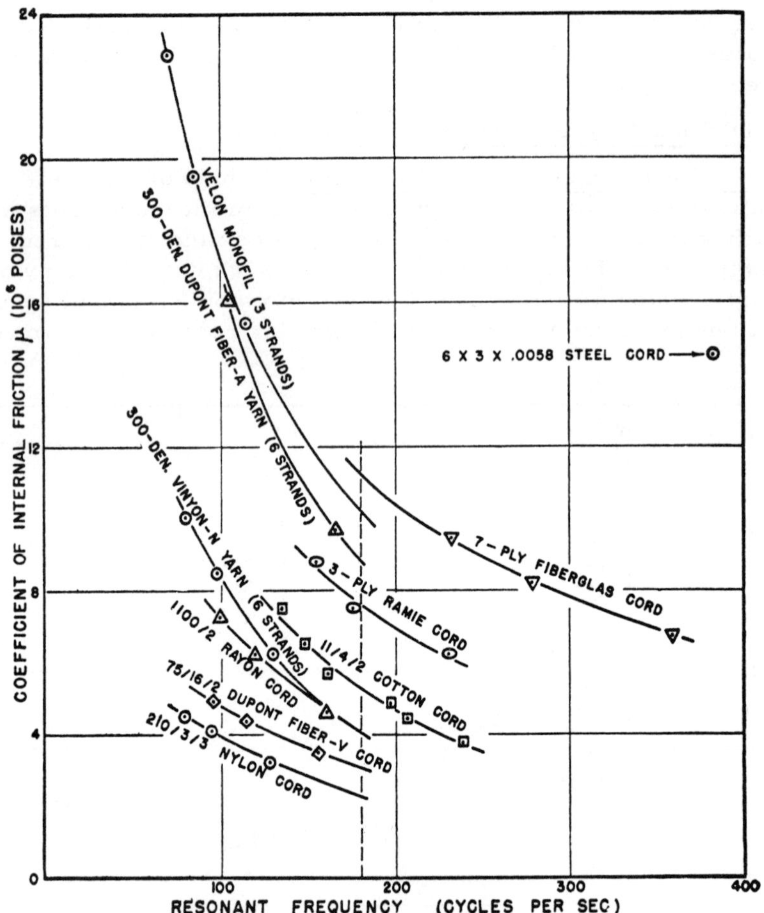

Fig. 5.8. Coefficients of internal friction of various yarns and cords as functions of frequency.[50]

That the loss modulus is largely independent of frequency was observed by Tipton also. His results on a continuous-filament rayon yarn in a range of frequencies at various strain-amplitudes are shown in the lower part of Figure 5.3.

5.5.2 Hysteretic Properties of Various Materials. In Table 5.2 are listed the friction-dependent properties of 14 of the samples in Table 5.1, the samples being in the same order. The tabulated values of the coefficient μ for most samples were interpolated or extrapolated for 180 cycles from the curves plotted in Figures 5.7 and 5.8. Also shown are loss tangents, and dynamic energy losses at unit strain-amplitude and at unit stress-amplitude. It can be seen that — as would be expected from the differences between the formulas by which they are calculated — no simple correlations exist among the values of any of the energy-loss functions.

Table 5.2. Dynamic Hysteretic Properties of Cords, Yarns, and Monofils[49]

(At 180 cycles/sec; Temperature Range: 21° - 25°C)

Description of Sample		Internal Friction (10^6 poises)	Loss · Tangent $\mu\omega_0/E$	Energy Loss	
Material	Construction			At Unit Strain Amplitude (10^{10} ergs/ cycle/cm^3)	At Unit Stress Amplitude (10^{-12} ergs/ cycle/$cm.^3$ / (dynes/ $cm^2)^2$)
Steel	Monofil; 6-mil	6.2	0.0033	2.22	0.0048
Fiberglas	7-ply cord	11.3	.024	4.00	.14
Ramie	3-ply cord	7.5	.027	2.66	.26
Rayon	1100-den yarn (2 strands)	4.9	.028	1.74	.45
Fiber A(du Pont)	300-den yarn (6 strands)	8.9	.061	3.16	1.18
Rayon	1100/2 cord	4.0	.030	1.42	.62
Fiber V (du Pont)	75/16/2 cord	3.1	.032	1.10	.92
Vinyon N	300-den yarn; Type NORU (6 strands)	4.1	.044	1.46	1.30
Cotton	11/4/2 cord	5.2	.058	1.85	1.78
Nylon 66	210-den yarn (9 strands)	3.1	.036	1.10	1.15
Nylon 66	210/3/3 cord	2.3	.030	0.82	1.08
Nylon 66 (high-tenacity)	Monofil (2 strands)	2.3	.033	0.82	1.28
Velon	Monofil (3 strands)	10.0	.176	3.55	8.66
Nylon 66	Monofil; 12-mil (3 strands)	2.7	.051	0.96	2.67

5.5.3 Characteristics of the Loss Tangent.   The loss tangent
$\mu\omega_0/E$ would be expected to vary only slightly with frequency, for,
as has been pointed out in the foregoing paragraphs, $\mu\omega_0$ is a
constant or very nearly so, and E is sensibly independent of
frequency, other conditions being uniform. How well this expectation is met by typical results is shown in Table 5.3, in
which are given the spreads between extreme values in the loss
tangents of the 14 samples listed in Table 5.2.   The important
variable here was resonant frequency, though other experimental conditions, such as specimen length, were necessarily varied also.

Table 5.3. Extreme Spreads in Loss Tangent of Various
Cords, Yarns, and Monofils, Resulting from Variations in
Resonant Frequency[49]

Sample Description	Frequency Range (cycles/sec)	Range of Loss Tangent $\mu\omega_0/E$
Steel monofil; 6-mil	128 – 195	0.0027 – .0045
Fiberglas cord	232 – 359	.0262 – .0298
Ramie cord	154 – 231	.0250 – .0291
Rayon, 1100-den yarn	114 – 225	.0259 – .0303
Fiber A (du Pont) yarn	105 – 165	.0607 – .0647
Rayon, 1100/2 cord	99 – 160	.0297 – .0310
Fiber V (du Pont) cord	95 – 155	.0276 – .0298
Vinyon N yarn; Type NORU	81 – 129	.0470 – .0493
Cotton cord	68 – 309	.0540 – .0695
Nylon 66 yarn	100 – 170	.0344 – .0370
Nylon 66 cord	79 – 127	.0262 – .0287
Nylon 66 monofil (high-tenacity)	65 – 300	.0275 – .0340
Velon monofil	72 – 115	.1648 – .1708
Nylon 66 monofil; 12-mil	67 – 109	.0545 – .0563

It can be seen that the loss tangent, while not strictly constant,
tends to remain within a narrow range, as the frequency is varied over a range which might be as high as fourfold, as in the case
of the cotton cord. It was noted,[49] regarding these data, that the
extremes of the frequency range are not to be identified with those
of the loss-tangent range; that is, they were not necessarily obtained in the same experiment. Hence, no systematic increase
or decrease of the loss tangent with frequency is to be deduced
from these data.

The manner in which the loss tangent is influenced by strain-
amplitude is shown in the lower part of Figure 5.6. It can be
seen that the loss tangent increases with increasing strain-
amplitude, and most markedly in the cabled yarn. Tipton as-

Table 5.4. Dynamic Properties of Widely Diverse Filamentous Materials[49]

Description of Sample	Dynamic Stretch Modulus (10^{10} dynes/cm^2)	Coefficient of Internal Friction (10^6 poises)	Frequency (cycles/sec)	Energy Loss	
				At Unit Strain-Amplitude (10^{10} ergs/ cycle/cm^3)	At Unit Stress-Amplitude (10^{-12} ergs/ cycle/cm^3/ (dynes/cm^2)^2)
Steel music-wire monofil; 6-mil (0.0152 cm) dia.,0.00018-cm^2 cross section	226	10.0	128	2.84	0.0056
Viscose rayon; 1100-den yarn; 2 strands; 0.0016-cm^2 cross section	19.7	4.9	180	1.74	0.45
Silk sewing thread; approx. 210 den	15.6	25.7	185	1.50	0.62
Wool yarn; 100%; 8-ply; approx. 4100 grex	5.2	21.6	110	0.74	2.76
Natural rubber golf-ball thread; 900% elongation	2.12	0.74	65	0.09	2.10

cribes this trend to the same interfiber mechanism that is thought to account for the decrease in the dynamic modulus, namely, the greater relative motion of adjacent fiber segments at the higher amplitudes. Since surface friction is involved, this mechanism provides for greater energy dissipation, which is reflected in the increase in the loss tangent. This interfiber movement would be expected to be most pronounced in the cabled yarn; and consistently, there is a much greater increase in loss tangent with strain-amplitude in the cabled yarn sample. Interfiber slippage, however, is not to be considered the sole cause of energy dissipation. In the low-twist, continuous-filament yarn, where relative interfiber movement is minimal, internal friction (at the submicroscopic level) undoubtedly plays an important role.

Figure 5.6 reveals also that the loss tangent decreases with increasing mean strain, at any given strain-amplitude. This behavior is the opposite of that of the dynamic modulus, as it is with respect to increasing strain-amplitude. In fact, the change in the loss tangent $\mu\omega_0/E$ with mean strain appears to be largely due to the change in E, with $\mu\omega_0$ remaining substantially constant.

5.5.4 Comparisons with Other Linear Structures.   The use of the stretch-vibrometer to measure the hysteretic properties of materials not of the usual textile type has been reported by Lyons. Dynamic properties of steel wire and cord, and glass-fiber cord are given in Tables 5.1 and 5.2.   The method was also used for measurements on a material exhibiting superelasticity. In Table 5.4 are given results on a rubber thread of the type used to form golf balls, together with those on some typical textiles and on a steel wire, which have a range of values for dynamic properties. It may be noted that as one goes through the list of materials from steel, in the order of declining dynamic modulus, there is a marked and steady decline in the energy losses at unit strain-amplitude, but, in general, an upward trend in the energy losses at unit stress-amplitude.

Chapter 6

MECHANICAL BEHAVIOR UNDER IMPACT LOADING

6.1 Introduction

Load-extension behavior and rupture properties associated with impact have been studied in a variety of textiles, ranging from single fibers to complex fabric structures. Most measurements have been made on linear structures subjected to longitudinal strain, but as has been brought out in preceding chapters, some observations have been made under transverse impact. A few typical results are given in Chapter 4. The following sections are devoted to a more thorough review of the findings on the impact behavior of various textiles. No attempt has been made for an uncritical tabulation in historical sequence of all or most results that have been obtained. Rather the plan has been to cite data and observations representative of the performance of textiles under particular conditions. The emphasis has been placed on the more recent definitive results, and those having more general interest and significance.

6.2 Linear Structures

6.2.1 Single Fibers and Monofils.
A few impact measurements have been made on single filaments. Hall[35] has studied the effect of strain rate on the stress-strain curve of oriented isotactic polypropylene, in connection with low-temperature tensile tests. His sample was a monofil produced with a draw ratio of about 5.5/1, and having a linear density of about 8.1 tex (73 den). In measurements at room temperature (20°C) he employed strain rates ranging from 3.3×10^{-4} to 4.9×10^{2} sec, using the apparatus of both Meredith[63] and Holden,[37] as well as conventional testers.

Hall's stress-extension results are shown in Figure 6.1. It may be noted in this graph that as the strain rate is increased the curves in any range of extensions move systematically to higher levels of stress. The breaking stress thus is found to increase with strain rate in most cases, for while there is a decline in breaking extension, it is insufficient to offset the higher stress levels of the more rapid tests. Breaking extensions fall into two groups: above approximately 40% for the low strain rates, and approximately 15 to 20% for the high rates. It is noticeable that the postyield inflection in the curves for the low, quasi-static rates ($< 3.3 \times 10^{-2}$ sec) disappears from the high-

121

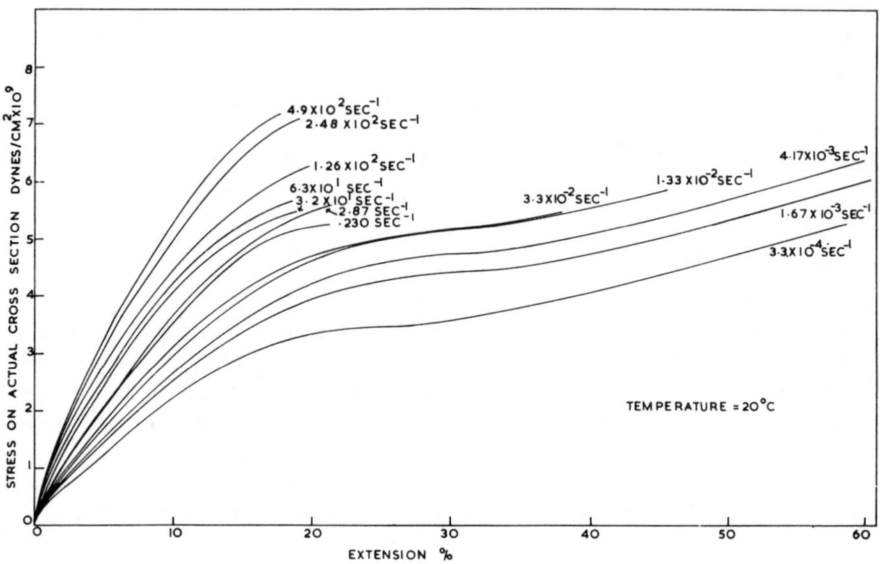

Fig. 6.1. Stress-strain curves of isotactic polypropylene monofils, at various strain rates.[35]

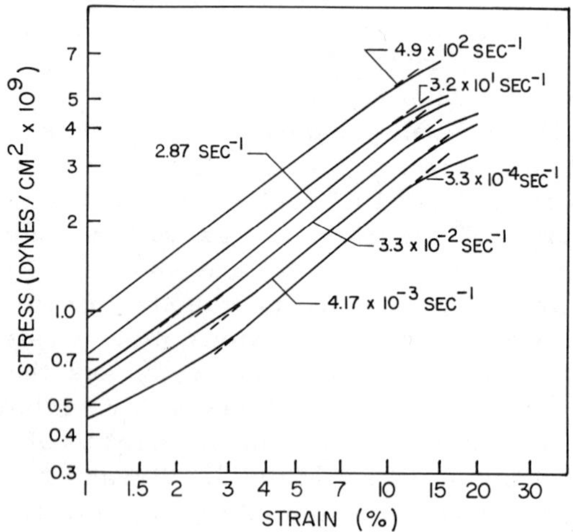

Fig. 6.2. Logarithmic plot of stress-strain relationships of isotactic polypropylene monofils, at selected strain rates. [After Reference 35.]

rate curves. At the high strain rates the monofils reach their
rupture point so rapidly that there is no time for the occurrence
of the plastic flow that accounts for the inflection regions.
On some semitheoretical considerations, Hall presents the
following equation:

$$\log \sigma = \frac{1-n}{m} \log \epsilon + \frac{n}{m} \log \dot{\epsilon} - \frac{1}{m} \log E \qquad (6.1)$$

where m and n are empirical, material constants and $\dot{\epsilon}$ is the
strain rate. On the basis of this equation, Hall concluded that if
the logarithm of the stress σ is plotted against the logarithm of
the strain ϵ, a series of straight lines whose slope is independent of the strain rate should be obtained.   Some of his results,
plotted in this manner, are shown in Figure 6.2. It can be seen
that between 1 and 10% strain at high strain rates, and between
3 and about 12% strain at low strain rates the curves are linear,
with substantially the same slope for all strain rates, as the
theory would predict.
Equation 6.1 also predicts that if log σ is plotted against log $\dot{\epsilon}$
for constant values of ϵ, a series of straight lines whose slope
is independent of ϵ should be obtained.   Hall's results, plotted
in this manner, for certain selected strains, are shown in Figure 6.3. The lines are not straight, but at any given strain rate
their slopes are quite uniform over the 0-to-14% range of strains.

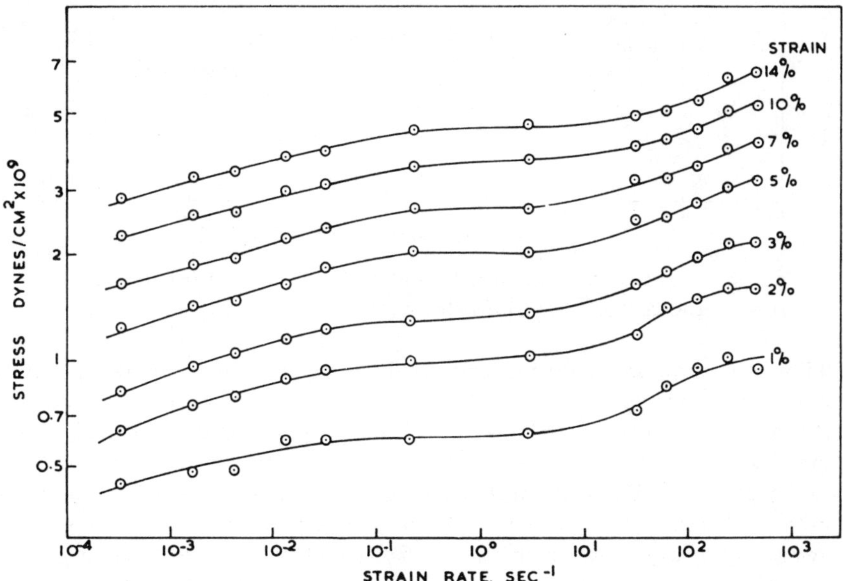

Fig. 6.3. Logarithmic plot of stress as a function of strain
rate in isotactic polypropylene monofils, at selected strains.[35]

On the basis of these results, Hall concluded that Equation 6.1 should be modified to read:

$$\log \sigma = \frac{1-n}{m} \log \epsilon + \frac{n}{m} \log \left[f(\dot{\epsilon}) \right] - \frac{1}{m} \log E \qquad (6.2)$$

in which the function $f(\dot{\epsilon})$ replaces $\dot{\epsilon}$ itself.

In Figure 6.4 the breaking parameters of the polypropylene are plotted as functions of the strain rate.   The energy to rupture was obtained by integration of the load-extension curve, using a planimeter. Also shown in Figure 6.4 are the stresses and extensions at the yield point.   It can be seen that at strain rates above about 30 %/sec the yield and breaking points merge, reflecting the fact, already noted, that the specimens break before they can yield. Further, regarding these results, Hall states:

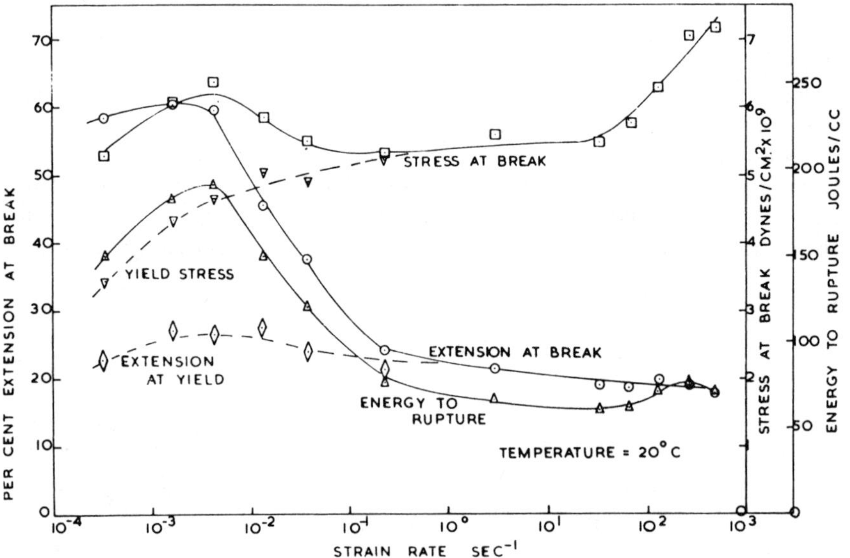

Fig. 6.4. Influence of strain rate on the breaking parameters of isotactic polypropylene.[35]

"The changes in breaking and yield stress with increasing strain rate are somewhat different.   As the strain rate is increased from 3×10^{-4} to 4×10^{-3} sec.[-1] [0.03 %/sec to 0.4 %/sec] the yield and breaking stresses increase, but the rate of increase of yield stress falls and the breaking stress decreases as the strain rate is further increased." As has been noted, the curves converge upon each other in the 10 %/sec region. The breaking stress remains constant with further increases in strain rate up to about 3000 %/sec, where a distinct rise in breaking stress occurs.

A group working at the laboratories of the Celanese Corpora-

tion has reported high-speed impact results on single fibers of
a number of synthetic materials.*   In Figure 6.5 are shown
tenacity-extension curves of Orlon 42 acrylic fibers, at seven
strain rates ranging from the quasi-static (10 %/min) to 1.8×10^6
%/min.   There is a consistent increase in the yield tenacity or
stress with increasing strain rate, as was found by Hall in polypropylene (Figure 6.4).   In the Orlon 42 there is also an increase
in breaking tenacity with increasing testing speed.   However, in
this fiber there appears to be no systematic dependence of breaking extension, nor of initial modulus and yield extension on the
strain rate.

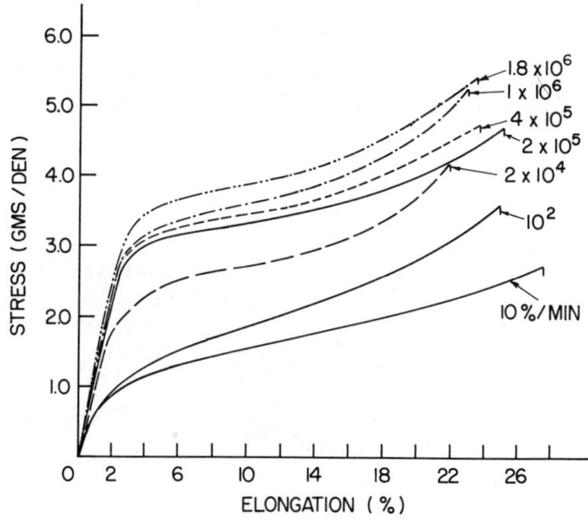

Fig. 6.5. Tenacity (stress)-strain curves of Orlon 42 acrylic
fibers at various strain rates. [After Reference 70.]

The Celanese workers were interested in obtaining some evi-
dence about the possible embrittlement of fibers at high rates of
loading. A measure of embrittlement was considered to be the
loss of ductility, that is, the ability of the specimen to stretch in
the postyield region, before rupturing. However, as Figure 6.5
shows, the Orlon 42 exhibited no such loss as a consistent effect
of increased strain rate. Further examples of this behavior are
shown in Figure 6.6 for acetate, and in Figure 6.7 for high-tenacity
viscose fibers.

While the increase in yield tenacity with increasing strain rate,
found by this group, agrees with Hall's results for polypropylene,
the latter's results, as given in Figure 6.1 do show a marked loss
of ductility at strain rates of 0.23/sec (1.4×10^3 %/min) or higher.

---

* Private communications. [Also, Reference 70.]

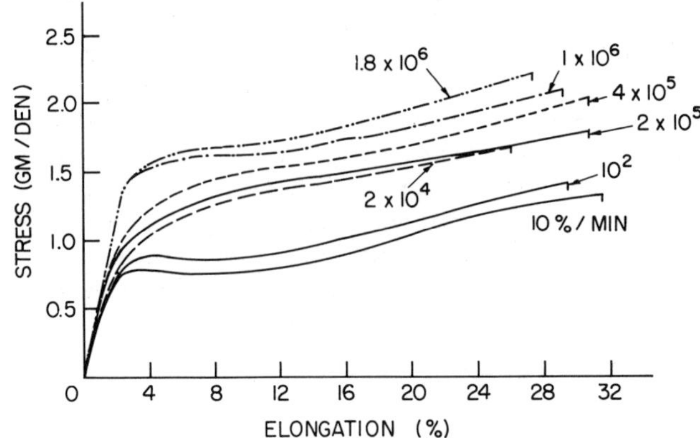

Fig. 6.6. Tenacity (stress)-strain curves of cellulose acetate fibers at various strain rates.   [After Reference 70.]

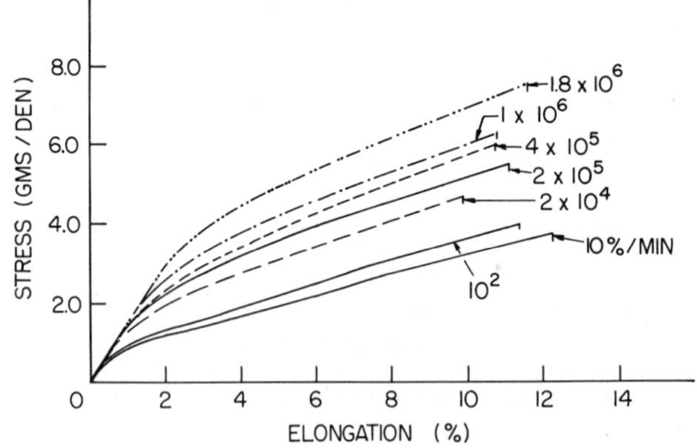

Fig. 6.7. Tenacity (stress)-strain curves of high-tenacity viscose fibers at various strain rates. [After Reference 70.]

The absence, at high strain rates, of the flow or relaxation that is manifested in ductility was also found by the Celanese group in another polyolefin, polyethylene. These results are shown in Figure 6.8. At a strain rate of 10 %/min, this sample yields at about 6 g/den, permitting such flow to occur that a 28% extension is obtained before the specimen breaks. Some ductility remains at 100 %/min. However, the molecular relaxation processes are relatively so slow in this material that at some strain rate, between 100 and 2×10^4 %/min, the available time intervals become too short for effective occurrence of the processes, so their influence on the behavior of the material is absent. Hence, at the high strain rates, very little yielding before rupture is seen in Figure 6.8.

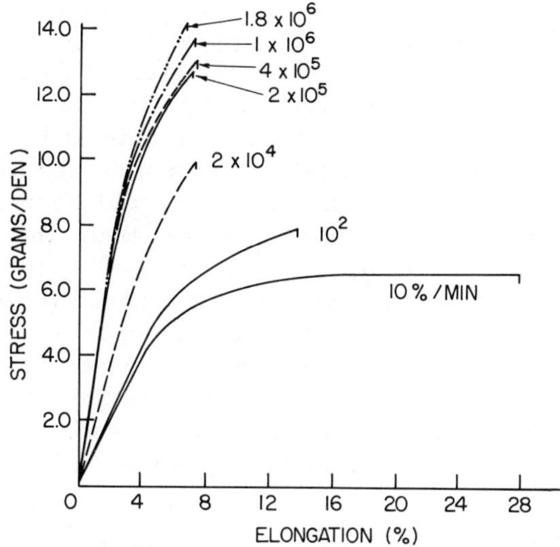

Fig. 6.8. Tenacity (stress)-strain curves of polyethylene fibers at various strain rates. [After Reference 70.]

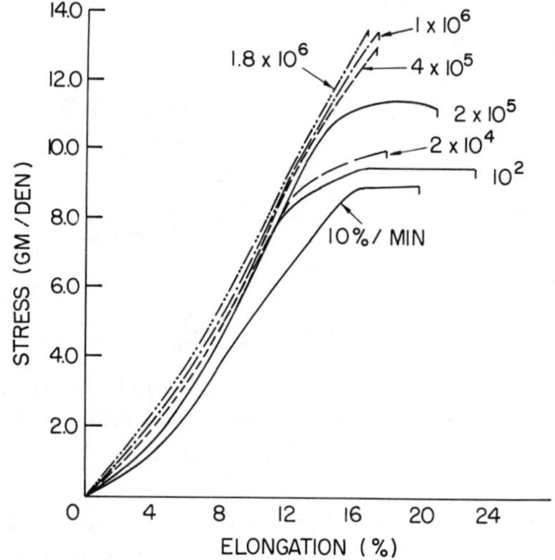

Fig. 6.9. Tenacity (stress)-strain curves of nylon 66 fibers at various strain rates. [After Reference 70.]

A similar situation was found to prevail in nylon 66 filaments, as is shown in Figure 6.9. Here, however, there is an incipient yielding at the 2×10^4 %/min strain rate.

The Celanese group speculated that some light might be thrown on the structural dependence of brittleness by an analysis of the times required for the various relaxation processes within a fiber.

To lead into this analysis, the tenacities at various extensions, as functions of the time required to reach the particular extensions, were desired. The theory underlying this treatment requires a constant rate of strain, such as is obtainable with the Celanese device. Since this strain rate was known in each of the tests, the time required (after the instant at which deformation started) to reach a given extension could be readily calculated. From a tenacity-extension curve, then, the corresponding tenacity could be found. It was thus possible to plot the tenacity at any selected strain as a function of the time required for that tenacity to be reached in the rupture process. The results on the polyethylene sample, for 3 and 6% extentions, are shown in Figure 6.10. These curves are interpretable as representations of the decay or relaxation of tenacity with increasing time, in a specimen held at a constant extension.

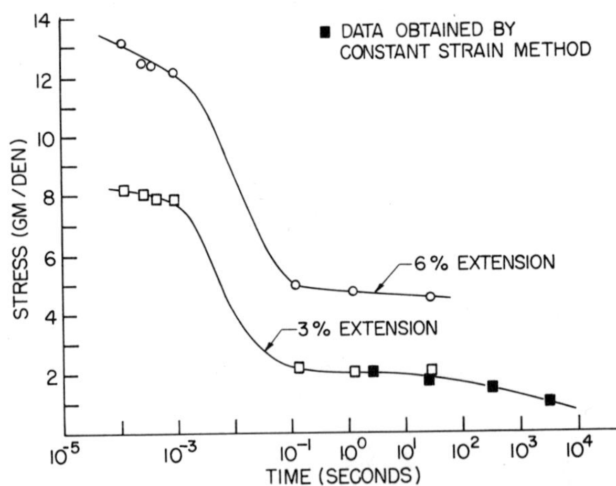

Fig. 6.10. Tenacity (stress) in polyethylene fibers, at two extensions (from Figure 6.8), as a function of time to reach these extensions. [After Reference 70.]

An auxiliary experiment was conducted, in which the polyethylene sample was stretched to a 3% extension and held there. The tensile force was then recorded as a function of time. The use of this method expanded the time range a hundredfold; the points obtained are indicated by solid-black squares in Figure 6.10. It can be seen that this line does indeed appear to merge with the 3% extension curve obtained by the impact method.

Figure 6.10 indicates that a major relaxation process occurs in this sample in the time range of 10^{-3} to 10^{-1} sec, at both levels of extension. However, significant relaxation occurs also in the short time intervals (10^{-4} to 10^{-3} sec) at the 6% extension, whereas

it is practically absent in this time range at the lower 3% extension. The Celanese group have concluded therefore, "that additional molecular mechanisms, with different relaxation times,
are involved at the higher strain levels."

6.2.2 Yarns and Cords. By far the largest number and widest
variety of impact experiments have been performed on yarns. A
brief look at Lester's[47] limited results will be of interest in showing that the distinction between impact and quasi-static values
was recognized even in this early work. His reported tests with
a falling pendulum were on a 75/6 unbleached sewing cotton (having an average breaking load of 420 g) by a conventional low-speed
method, using a 50-cm gage length.   From the load-extension
diagram Lester calculated an average energy to rupture of 1882
g-cm. In impact tests on the sample, using the same 50-cm
length, he found an average energy to rupture of 1995 g-cm.
Finding the impact energy to rupture to be greater than the quasi-
static presaged many similar results that were to be obtained in
the following four decades.

Lester also tested his sample on the impact instrument using
a 25-cm gage length, half that mentioned in the previous paragraph. He found a rupture energy of 1204 g-cm, and noted that,
if the rupture energy were strictly proportional to the length of
the specimens, he should have obtained 997 g-cm.   The reason
for the discrepancy, though it was not recognized by Lester, is
that in the longer gage length there is greater probability of finding weaker flaws, a concept in extreme-value statistics.

Using an impact pendulum, Denham and Brash[25] found a similar dependence of rupture energy on specimen length in an undegummed silk yarn (2/18 den).   They tested their sample at 4
lengths, using a bundle of 2 or 3 specimens in each test. Thirty
specimens were broken at each length. The results are shown in
Table 6.1. It can be seen that the breaking energies per cm for
the 1-, 2-, 3-, and 4-cm lengths are 35.8, 33.2, 32.3, and 31.8
g-cm, respectively; there is a consistent decline as the testing
length is increased.

Table 6.1. Breaking Energy of Silk Yarn at Different Lengths[25]

Number of Yarns in Bundle	Length of Yarns (cm)	Mean Breaking Energy/Yarn (g-cm)
3	1	35.8
3	2	66.4
3	3	97.0
2	4	127.4

Recognizing the regularity of this decline in energy/unit length, Denham and Brash derived the following simple equation to express the dependence of the rupture energy A_r on the initial length ℓ_0 of the specimens:

$$A_r = a\ell_0 + b \tag{6.3}$$

where the constants a and b are 30.6 g and 5.2 g-cm, respectively, for this particular sample. Equation 6.3 gives the rupture energy for each testing length, the following:

1 cm	35.8 g-cm
2 cm	66.4 g-cm
3 cm	97.0 g-cm
4 cm	127.6 g-cm

These will be seen to be in extraordinary agreement with the energies listed in Table 6.1.

Denham and Brash appear to be perhaps the first to inquire into the influence of impact velocity on the breaking energy of textiles. They broke 2-cm specimens of the undegummed silk yarn by allowing the pendulum to fall from three different heights, to obtain three impact velocities. A large number of determinations were made at each condition. The velocities of impact (at the center of percussion) were checked by allowing the pendulum to fall with no specimen in place. The results of these experiments are summarized in Table 6.2. The data displayed therein tend to support these workers' conclusion that a "comparison of

Table 6.2. Breaking Energy of Silk Yarn at Different Impact Velocities[25]

Number of Yarns in Each Test	Total Number of Yarns Broken	Mean Breaking Energy/Yarn (g-cm)	Impact Velocity (cm/sec)
1	90	66.5	82.1
2	90	66.9	82.1
1	90	65.8	119.0
2	90	66.5	119.0
3	810	66.4	119.0
4	92	67.5	119.0
3	270	67.5	137.6
4	92	69.7	137.6

these mean values indicates an increased absorption of energy
with increasing velocity of the pendulum at the time of break,
the mass of the pendulum and of the rider [on the scale] remaining constant." The effect, however, is so small as hardly to be
significant.

These authors made comparisons of their impact results with
those obtained from stress-strain diagrams for quasi-static tests,
but presented no data. They concluded, however, "that the work
done in breaking by a slowly increasing load is about one-half only
of the work done in breaking by the pendulum."

While Midgley and Peirce[67,75] thoroughly analyzed the performance of the impact pendulum and other tensile-testing equipment,
they did not publish many results of general interest. They did,
however, present some data related to the question discussed in
this section; but, rather than consider impact velocity directly,
they focussed their attention on the speed with which the rupture
was accomplished (the rupture time-interval). In separate experiments they varied the speed in two ways: "by breaking similar specimens with different heights of fall, or by breaking different numbers of threads with the same fall." By the first
method, the rupture interval would be altered because the velocities at the instant of impact would be changed. In the second
method, the speed of rupture is controlled by the resistance of-
fered by the specimens.

In both experiments the sample was a No. 36 Sakel cotton yarn
of good regularity. For the first method, 54 specimens of 30
yarns each, 24 in. long, were tested at each height of fall. The
latter quantity was reported in arbitrary units related to the geometry of the machine; a larger number of units represents a
greater height of drop. The results of this experiment are given
in Table 6.3.

In the second experiment, 20 specimens of each bundle size
(number of yarns), 67.4 cm long, were tested at a constant height

Table 6.3. Breaking Energy of Cotton Yarn Broken by Pendulum Released from Various Heights[67]

Height of Drop (arbitrary units)	Rupture Interval (sec)	Breaking Energy (g-cm/cm)
500	0.034	15.5
700	.024	15.8
900	.019	15.5
1100	.017	16.1

of fall of 1100 units. The data obtained employing this method are summarized in Table 6.4. The rupture intervals entered in Tables 6.3 and 6.4 were calculated according to an equation derived by Peirce,[75] involving the breaking extension of a specimen and the kinetic energy of the pendulum at any instant. This calculation was apparently based on the assumption (now recognized as unwarranted) that the breaking extension is the same in impact loading as in a quasi-static test. Accordingly, the rupture intervals given in the tables can, at best, be taken only as relative.

Table 6.4. Breaking Energy of Cotton Yarn Broken in Bundles of Various Numbers of Yarns, with Constant Height of Drop[67]

Number of Yarns in Bundle	Rupture Interval (sec)	Breaking Energy (g-cm/cm)
80	0.038	15.7
60	.026	15.8
40	.021	16.0
20	.018	16.3
10	.016	17.0

These data show a trend that is in agreement with the results of Denham and Brash, that is, an increased absorption of energy (breaking energy) as the speed of the rupture process is increased. Midgley and Peirce recognized that the differences were, however, too small to permit estimation of the relationship in quantitative fashion.

The latter authors attempted to calculate breaking loads from the energies given in Tables 6.3 and 6.4, again on the erroneous assumption that breaking extensions are constant over all rates of loading. It was not until electronic devices were adapted to impact testing that load-extension relationships during the impact rupture process could be followed directly. What is perhaps the earliest published diagram showing such a relationship is shown in Figure 6.11,[90] in comparison with that for the quasi-static loading of the same sample. These curves were obtained on nylon 66 (supposedly a heavy continuous-filament yarn or cord) with the M.I.T. equipment described in Chapter 3.[36,89]   The dynamic (impact) rate of loading of 21 ft/sec is 1200 times greater than that of the quasi-static test, 12 in./min. In these curves can be seen the tendency for the breaking load to increase, and the breaking extension to decrease, at high strain rates, as was later found for single fibers of some other man-made textiles in the work of Hall and the Celanese group, cited in Section 6.2.1.

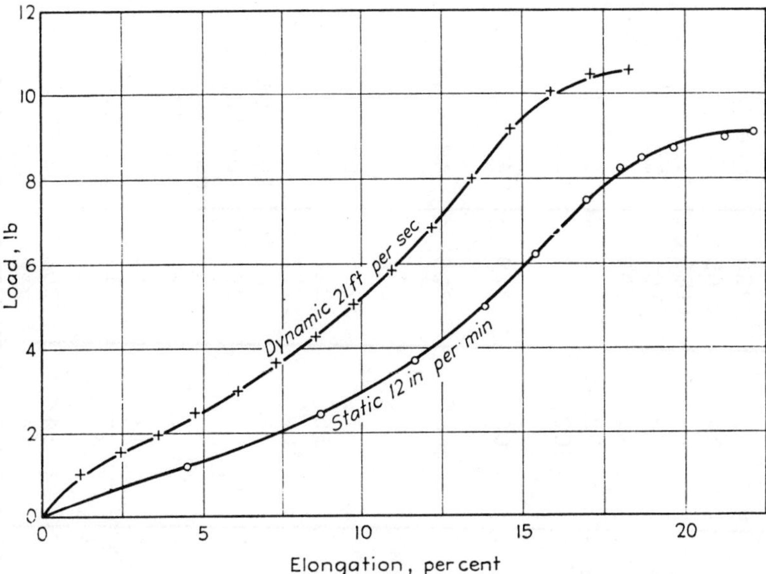

Fig. 6.11. Effect of strain rate on load-extension behavior of nylon 66 yarn.[90]

In their work employing a falling-weight apparatus, Schiefer and associates[86] determined the impact rupture energy of a wide variety of threads, yarns, and monofils.   Their procedure, as described in Chapter 3, gave them (in effect) the kinetic energy of the falling weight at which 50% of the specimens of a given sample could be expected to break.   For comparison with the impact results, they also obtained low-speed load-extension data on the sample, and from these computed quasi-static rupture energies.   The samples tested are listed in Table 6.5.   There is some duplication, but all the data have been reproduced here, since this will provide the reader with some insight into reproducibility and the influence of minor differences.

The results of the quasi-static and impact tests are given in Table 6.6.   Some specimens of each sample were loaded up to only 80% of their average breaking force in the low-speed tensile tester, and then were completely unloaded. Force-extension curves were obtained for both halves of the cycle.   Measurement of the area under the loading curve gave the work done on, or energy put into, the specimen; while the area under the unloading curve gave the energy recovered from the specimen in this part of the cycle.   The latter area was then expressed as a fraction of the input area, in percentage, to give the energy recovery, values of which, for each sample, are given in Table 6.6. Rupture energy is given in the table as breaking toughness, that is, as energy per unit length per unit linear density (see Chapter 2).

Table 6.5. Materials Used in Falling-Weight Impact Studies[86]

Sample Number	Description	Number of Filaments	Linear Density (den)	Twist (t.p.i.)
1	Polyethylene, undrawn	1	3820	none
2	Nylon 66, undrawn, treated	34	2165	1Z
3	Nylon 66, undrawn	17	1185	1Z
4	Nylon 66, undrawn, treated	34	1163	1Z
5	Nylon 66, undrawn, treated	34	1195	1Z
6	Acetate, high impact	52	343	3S
7	Vinyl chloride-acrylonitrile copolymer, 70% elongation	200	407	12Z
8	Vinyl chloride-acrylonitrile copolymer, 50% elongation	200	329	4Z
9	Vinyl chloride-acrylonitrile copolymer, 40% elongation	200	305	4Z
10	Vinylidene chloride-vinyl chloride copolymer	1	840	none
11	Vinylidene chloride-vinyl chloride copolymer	1	1235	none
12	Vinylidene chloride-vinyl chloride copolymer	16	220	6Z
13	Rayon, viscose	480	1135	3Z
14	Acrylonitrile, 3-strand sewing thread	1	945	20S(10Z)*
15	Nylon 66, 3-strand sewing thread	1	1060	0(10Z)*
16	Nylon 66	1	318	none
17	Silk	–	788	12S(9Z)*
18	Nylon 66	17	258	1Z
19	Rayon viscose, high strength, 3-strand sewing thread	–	917	14S(12Z)*
20	Polyester	34	206	2.5Z
21	Nylon 66	34	208	1Z

*   Values in parentheses represent cord twist.

Table 6.6. Rupture Properties of Textile Materials Listed in Table 6.5

Sample Number	Quasi-Static Properties				Impact Breaking Toughness (in-lb/in-den)
	Breaking Tenacity (gm/den)	Breaking Extension (%)	Energy Recovery (%)	Breaking Toughness (in-lb/in-den)	
1	0.3	750	3	26×10^{-4}	27.5×10^{-4}
2	0.6	550	5	41	37.0
3	0.7	570	8	65	49.5
4	0.8	420	9	57	74
5	1.1	420	7	58	61
6	1.1	37	6	6.5	10.7
7	1.7	66	9	16.2	18.0
8	2.0	45	12	11.5	15.5
9	2.3	38	13	10.9	14.7
10	1.9	26	46	6.7	7.5
11	2.0	24	40	5.6	6.8
12	2.0	13	73	2.5	5.6
13	3.0	9	32	3.8	6.8
14	3.0	17	28	6.1	14.5
15	3.5	20	75	6.9	12.5
16	3.9	13	67	4.9	15.5
17	5.2	15	33	10.9	15.0
18	5.9	14	59	8.2	10.3
19	6.0	8	37	5.0	9.8
20	6.6	12	60	10.2	15.3
21	6.7	16	63	13.4	16.5

Concerning these results, Schiefer and colleagues noted that they reflect basic differences, not masked by the variety in number of filaments, linear density, and twist of the samples: "Thus the energies, both [quasi-] static and impact, to break the yarns of low [breaking] tenacity and very high elongation were relatively high. The recovered energies for the yarns of high and very high [breaking] tenacity and ordinary [breaking] elongation ranged from moderately high to very high. . . . The impact energies to rupture of the yarns of very high elongation, nos. 1, 2, 3, 4, and 5, are about equal to their corresponding static energies. On the other hand, the yarns having high or very high [breaking] tenacity and ordinary [breaking] elongations, nos. 12, 13, 14, 15, 16, 17, 19, and 20, have values for impact energy to rupture which are considerably greater than the static energies."

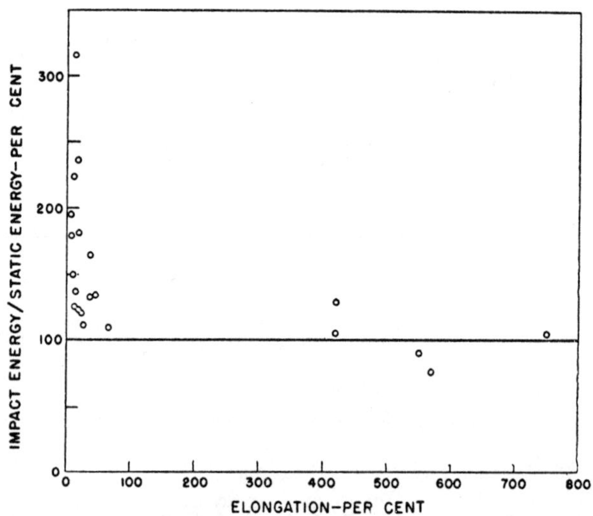

Fig. 6.12. Ratio of impact rupture energy to quasi-static energy of various samples, plotted at corresponding quasi-static breaking extensions.[86]

In Figure 6.12 the ratio of the impact rupture energy to the quasi-static energy of each sample, expressed as a percentage, is plotted against the quasi-static breaking extension (elongation). It can be seen that there is a tendency for this ratio to become large as the quasi-static breaking extension decreases, especially below 50%. The plot brings out again what was noted in the previous paragraph, in slightly different language; namely, that where the sample has great ductility (is capable of considerable elongation), as are the undrawn samples nos. 1, 2, 3, 4, and 5, the rupture energies tend to be the same, whether the sample

is broken in impact or quasi-statically. No rigorous generalization, however, can be drawn from these data.   The Schiefer
team concludes: "There appears to be no general relationship
between the [breaking] tenacity and impact energy to rupture of
the yarns, or between the impact energy to rupture and [break-
ing] elongation." This is to say that the quasi-static tensile properties of textile samples provide no reliable bases for estimat-
ing their impact rupture energies.

The unpredictability not only of impact rupture energies, but of
impact breaking forces, from low-speed properties was brought
out in the work of Lyons and Prettyman on tirecords.[51]   They
used a falling pendulum that provided an initial impact velocity
of about 380 cm/sec ($\approx$ 9000 in./min), and, for the quasi-static
test, a constant-rate-of-traverse (Scott X-3), and a constant-

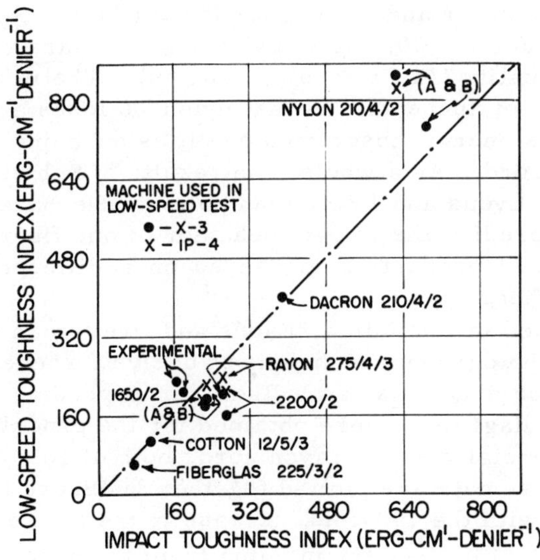

Fig. 6.13. Comparison of the impact and low-speed rupture
energies of various tirecords. The letters "A" and "B" refer to different samples of the same type of cord.   [After
Reference 51.]

rate-of-loading (Scott IP-4) machine. Their results, comparing
impact and quasi-static energies of rupture of a variety of tirecords, including one made from an experimental, saponified, acetate fiber, are shown in Figure 6.13.* The impact energies were

---

* The term "toughness index" is used in Figure 6.13 as an alternative for breaking toughness. The toughness indices in the
figure have the same dimensions (though not the same units) as
breaking toughness in Table 6.6. The unit "in.-lb/in.-den" =
4.45×10^5 erg/cm-den.

obtained from the usual readings on the pendulum scale, while the low-speed values were obtained by measurement of the areas under the load-elongation curves on tensilgrams.

Shown in Figure 6.13 as a broken line is the locus of equal quasi-static and impact values. It can be seen that there is a qualitative agreement between the two sets of experimental values in that the fiber types rank the same in the two sets. Furthermore, the points for three of the samples, Fiberglas, cotton, and Dacron, fall on the equal value line. However, it can be seen that the low-speed indices are much higher than the impact indices for the nylon 66 (about 35% in one case) and the experimental (about 30%) cords. For the rayon cords, on the other hand, the impact indices are the higher (by 20 to 25%, on the average). It may be noted that while these results show the nylon 66 cords to have greater quasi-static than impact breaking toughness, Schiefer and colleagues found (Table 6.6) the reverse for their sample no. 20, a low-twist nylon 66 yarn, presumably of tirecord quality (34 filaments, 208 den). The inference from a comparison of these data is that nylon 66 loses a substantial fraction of its energy-absorption qualities on being twisted and formed into cord. As a whole, the results of Schiefer and colleagues, and Lyons and Prettyman point up the hazards of attempting to predict the impact behavior of one fiber assembly from low-speed properties measured on a different assembly of the same fiber type.

As described in Chapter 3, Lyons and Prettyman employed a high-speed photographic technique to derive force-extension diagrams. Such diagrams, as well as corresponding low-speed (12 in./min) diagrams, were obtained for the experimental cord and a commercial viscose rayon cord, both of 1650/2 construction. All these data are plotted together in Figure 6.14. It can be seen that with the 750-fold increase in the rate of elongation in going from the quasi-static to the impact test, the viscose control exhibits a slight increase in the breaking force, and a decrease in the breaking extension. Such alterations in force-extension curves for comparable changes in strain rates have

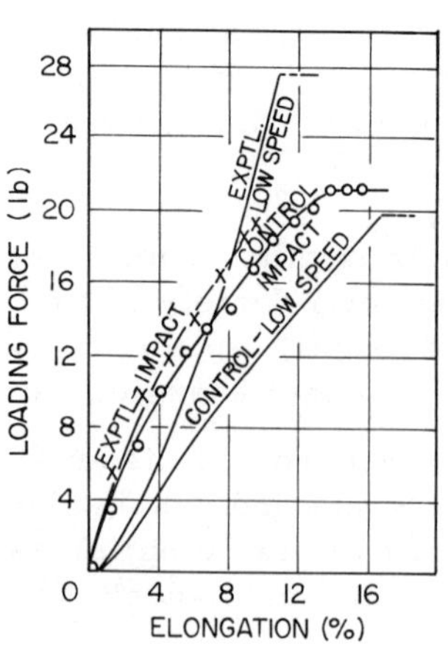

Fig. 6.14. Comparison of force-extension curves of two cellulosic tirecords obtained in impact and low-speed tests. [After Reference 51.]

been noted in results cited previously, as for example, those
shown in Figure 6.11.   In the experimental sample, however,
there is a 30% drop in breaking force between the low-speed and
the impact tests. In fact, in this sample the whole character of
the tensile curve is changed: From being concave-upward in the
quasi-static test it becomes concave-downward in the impact test.

In most impact experiments, Schiefer and colleagues had a rate
of extension of about 1000 %/sec, though for some short specimens (samples nos. 1 through 5) they attained a rate of about
5000 %/sec. In Lyons and Prettyman's studies the impact strain
rates were about 750 %/sec. The influence on force-extension
relationships of rates intermediate between these and the quasi-
static was not studied by either of these teams. This was done
by Meredith[63] for a variety of yarns, at rates of extension ranging from 0.0007 up to over 1000 %/sec. By his method, which
employed a rotating disc for the high-speed tests, he was able
to obtain directly a photographic record of what was substantially
the force-extension diagram, for each specimen. Typical results
are shown in Figures 6.15, 6.16, and 6.17.*

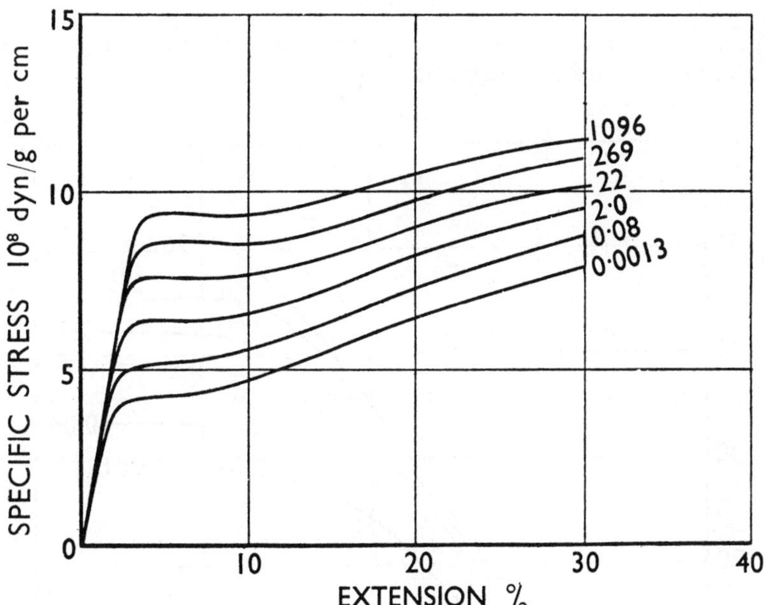

Fig. 6.15. Tenacity-extension curves of cellulose acetate
yarn at various rates of extension (in %/sec).[63]

---

*The "specific stress" appearing in Figures 6.15, 6.16, and
6.17 is the same quantity as tenacity, here expressed in "10^8
dynes/g per cm," instead of "g/den." 10^8 dynes/g per cm = 0.111
g/den = g/tex.

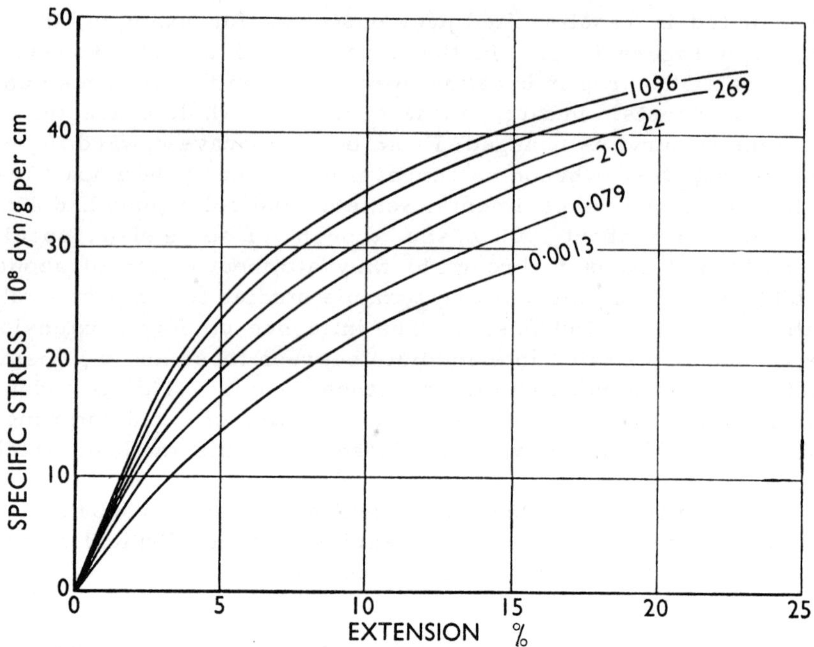

Fig. 6.16. Tenacity-extension curves of silk yarn at various rates of extension (in %/sec).[63]

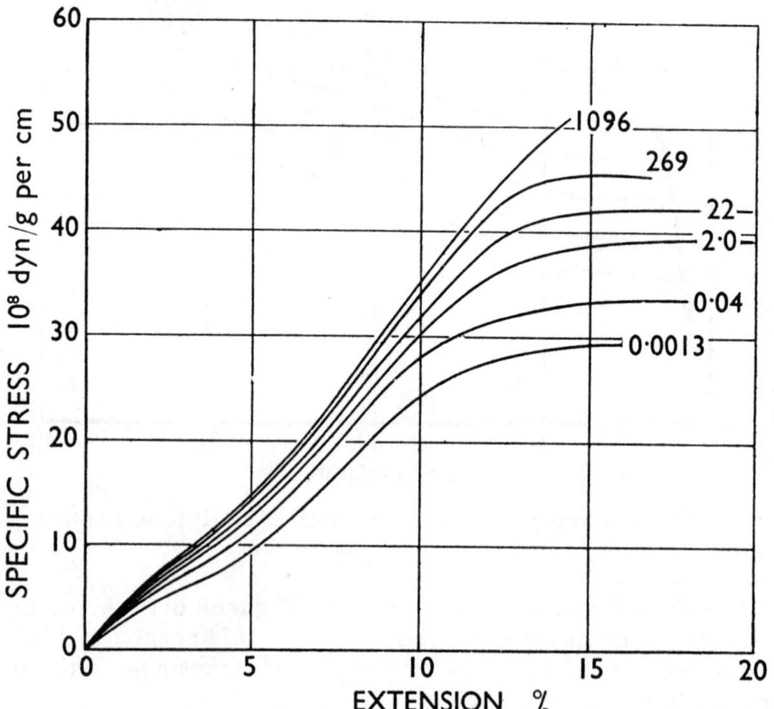

Fig. 6.17. Tenacity-extension curves of nylon 66 yarn at various rates of extension (in %/sec).[63]

Meredith summarizes his findings as follows: "At any given extension the stress is higher the higher the rate of extension and is approximately linearly related to the logarithm of the rate of extension for all the yarns except the cellulose acetate.   The strength [or breaking tenacity], and the modulus of elasticity for the initial linear part of the specific-stress vs strain curves, increases approximately linearly with the logarithm of the rate of extension.   The breaking extension. . . of cellulose acetate rayon is independent of the rate of extension.   Breaking extension increases considerably with rate of extension for the silk yarn, whereas the breaking extension of the nylon yarns show a maximum value for a rate of extension of about 2 %/sec."

How well the breaking tenacities of the silk and nylon 66 yarns vary linearly with the logarithm of the extension rate is shown in Figure 6.18, the data being estimated from Meredith's curves.

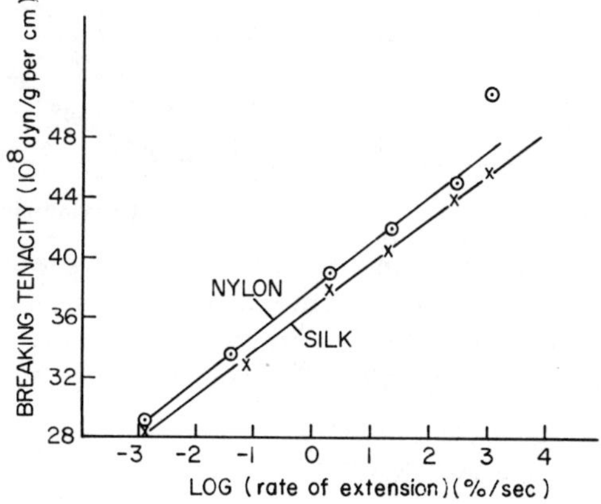

Fig. 6.18. Breaking tenacities of silk and nylon 66 yarns as functions of the logarithm of the rate of extension (estimated from Meredith's data in Figures 6.16 and 6.17).

The most extensive results on the impact properties of yarns and the influence of strain rate are those published by Schiefer, Smith, and associates over the past eight years.   In one of their early studies[87] they were concerned with comparing the force-extension behavior of "dry" and "wet" yarns at different strain rates, without rupturing the specimens.   The dry specimens were in equilibrium with atmospheric conditions of 50% relative humidity and 70°F, while the wet specimens had been immersed in water for 3 to 6 hr before testing. Curves obtained in the loading and unloading of high-tenacity nylon 66 yarns are shown in Figures 6.19 and 6.20. The curves for the dry condition show a greater slope for impact, than for quasi-static loading, with an even larger difference in the low-extension region than was found

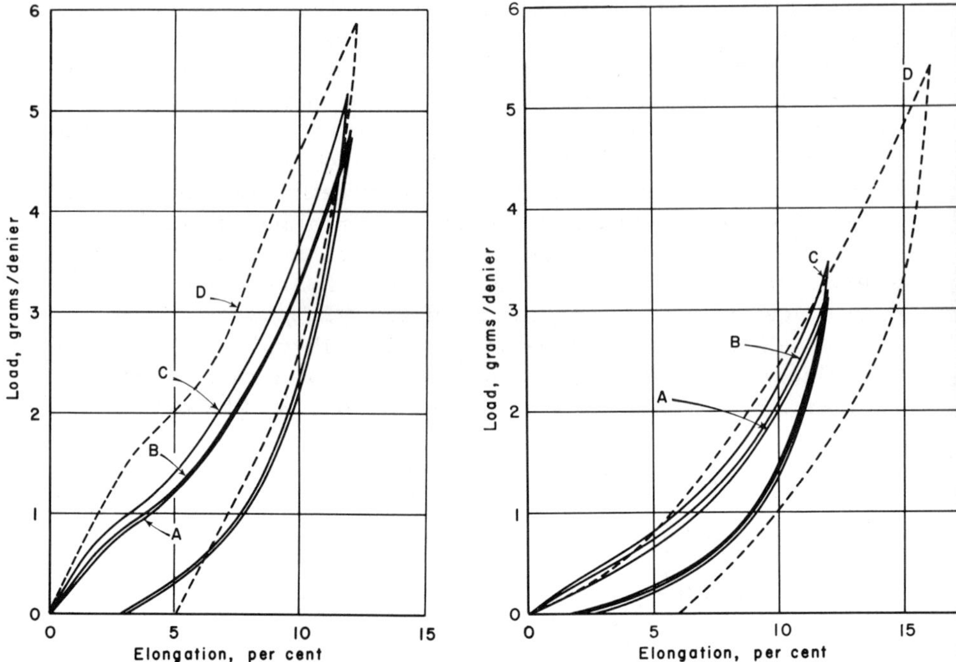

Tenacity (load)-extension curves for loading and unloading
of high-tenacity nylon 66 yarn. Curves A, B, and C = rates
of strain: 1, 10, and 100 %/min, respectively. Curve D =
longitudinal impact: 315,000 %/min initially, to 0 %/min at
maximum load.[87]

Fig. 6.19. In dry condition.   Fig. 6.20. In wet condition.

by Meredith (Figure 6.17). Comparing Figures 6.19 and 6.20:
Water in the sample appears to have a softening effect in the impact test, so that at a given level of tenacity (load), the extension is at least 50% higher than in the dry tests. Even more pronounced changes in tenacity-extension curves were found to result from the introduction of water into Fortisan and experimental
saponified acetate yarns.

Smith and associates have recently published the results of
their experimental survey of rupture properties as functions of
strain rate for 19 different yarn samples in 12 fiber types.[101] In
their quasi-static tests, they employed strain rates of 1, 10, and
100 %/min on all samples. For the determination of breaking
tenacity and extension under impact they used the transverse
method, providing strain rates on the order of 5000%, in most
cases. The longitudinal impact technique was also used, but
only to obtain, directly, breaking energy density. Representative

Table 6.7. Rupture Properties of Various Yarns in Quasi-Static and Transverse Impact Tests [101]

Fiber Type of Yarn	*	Initial Textile Modulus (g/tex)	Breaking Tenacity (g/tex)	Breaking Extension (%)	Breaking Energy Density (joules/g)
Cellulose acetate	A	370	11.2	32.0	25.8
	B	350	14.6	30.7	34.9
Cellulose triacetate	A	450	12.7	24.2	22.5
	B	400	16.8	15.6	18.9
Cotton (sewing thread)	A	570	28.0	6.4	9.7
	B	860	34.5	6.8	13.6
Polyester	A	1100	44.0	22.3	80.3
	B	1100	60.5	8.0	24.3
Glass fiber	A	2430	64.5	2.7	8.5
	B	2470	63.1	2.6	8.1
Human hair, 50 strands	A	380	14.4	36.7	36.6
	B	420	17.3	33.6	43.3
Vinal	A	940	58.3	15.1	41.0
	B	1130	68.9	14.3	48.2
Nylon 66, high-tenacity	A	480	75.8	15.5	55.4
	B	860	81.4	11.1	38.5
Acrylic	A	980	50.6	17.0	43.2
	B	1020	60.6	14.6	46.5
Rayon, low-tenacity	A	610	18.3	20.7	24.6
	B	590	22.8	19.3	28.3
Rayon, tire yarn	A	500	26.0	19.0	28.1
	B	650	38.8	19.1	44.8
Saran	A	130	18.7	18.3	16.9
	B	170	21.7	15.3	17.4
Silk, braided fish line	A	470	36.5	19.0	43.6
	B	490	42.3	17.9	46.4

*Method of testing: A = quasi-static at 100 %/min; B = transverse impact at about 40 m/sec (missile velocity).

results are presented in Table 6.7; the quasi-static data are those obtained at 100 %/min, while the impact values are those given by the transverse method. The breaking energy densities for both types of test were computed from the areas under the tenacity-extension curves.

Concerning features of the results common to all the fiber types, Smith and associates note that "curves obtained at the higher rates of straining have steeper initial slopes [higher moduli]. The initial linearity of these higher-rate curves tends to persist to higher values of stress. The breaking tenacities are usually higher, but the breaking elongations may be greater or less than the values obtained at conventional rates of testing, depending upon the material tested."

Tenacity-extension curves for all 19 samples were presented by Smith and associates. Their curves for cellulose acetate, triacetate, and human hair are very similar to those of Meredith for the first of these materials (Figure 6.15). Not unexpectedly, the curves for the glass-fiber yarn, which is not a viscoelastic material, are strictly linear, with little difference between those obtained in the quasi-static and in the impact tests. The curve for the behavior of high-tenacity nylon 66 yarn under transverse impact is very similar in shape to that of Meredith for longitudinal impact at 1096 %/sec (Figure 6.17), and the loading section of the impact curve of Schiefer and associates for the dry yarn (Figure 6.19). The Smith team's impact curve, however, shows about 100% higher tenacity for any given extension than do either of the earlier curves. The curves for the rayon samples, from low-, through medium-, to high-tenacity, and tirecord yarns all show the same shape: a more pronounced yield region and extended inflection region than appear in curves for nylon 66, for example. Typical curves for rayon are shown in Figure 6.21.

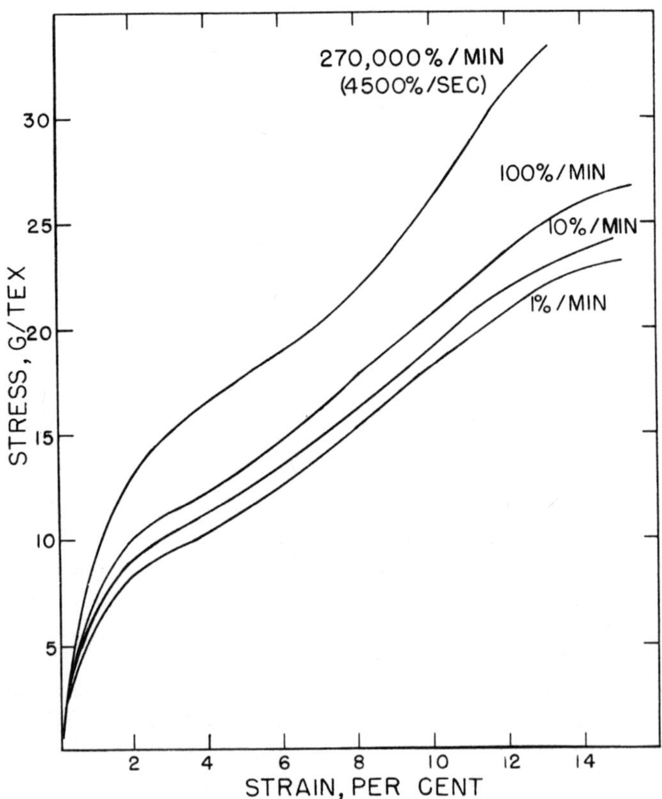

Fig. 6.21. Tenacity (stress)-extension curves of medium-tenacity rayon yarn at various rates of extension.[101]

With his ballistic method, Lewis[48] attained rates of strain in
the order of 23,000 %/sec, from 3 to 17 times the rates employed
by Smith and associates.   His results for a nylon 66 monofil,
continuous-filament cellulose acetate, and polyethylene tere-
phthalate (Terylene) yarns are shown in Figure 6.22.   It can be
seen that up to about 12% extension, the curve for nylon closely
resembles the loading curve of the same material found by
Schiefer and associates (Figure 6.19).   Lewis' curve shows a
second yield point, at about 12%, not generally found at very high
rates of strain, as (for example) in Meredith's work, shown in
Figure 6.17. The acetate yarn exhibits the same extended yield
region shown for this material in Figure 6.15, but the breaking
extension is nearly 50% less than that found by Meredith, as well
as by Smith and associates (Table 6.7). This reduction in breaking extension as the strain rate is greatly increased has been
noted in other textile samples, as, for instance, in connection
with Figure 6.1.

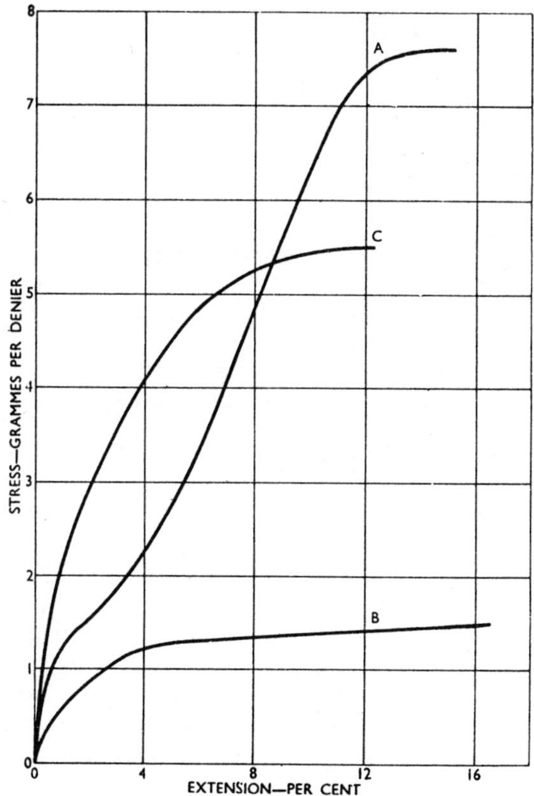

Fig. 6.22. Tenacity (stress)-extension curves of nylon monofils (A), cellulose acetate (B), and Terylene (C) yarns at
a strain-rate of about 23,000 %/sec.[48]

It is evidently not invariably true that the breaking extension decreases in all textile materials as the strain rate is increased indefinitely.   Using a modified version of Lewis' apparatus, Holden[37] investigated the effects of strain rates up to 66,000 %/sec in nylon 66 yarns of several types. He found the breaking strain of undrawn, annealed nylon to increase from about 16 to 29% as the strain rate was raised from 0.033 to 66,000 %/sec. His results for drawn nylon are shown in Figure 6.23, in which are also included the results of Meredith[63] and Smith, Schiefer, and colleagues.[100] At low strain rates there is an increase in breaking extension with increasing strain rate up to the neighborhood of 1 to 10%. As the strain rate is increased to the order of 10^3 %/sec the breaking extension drops, but passes through a minimum.

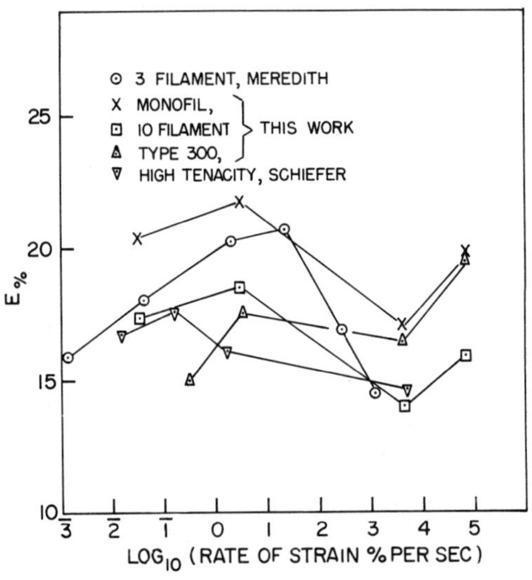

Fig. 6.23. Breaking extensions (E) of various drawn nylon 66 samples as functions of the logarithm of the rate of extension. [After Reference 37].

Figure 6.23 places the minimum at about 4000 %/sec.   Lewis' results on the drawn nylon monofil in Figure 6.22, where the breaking extension is shown to be about 15%, suggests that the minimum may be at a strain rate 5 times higher.   In any case, the results show the breaking extensions to rise when the strain rate is at 66,000 %/sec, in all three samples tested at that rate. Holden offers no explanation for this apparent, more or less anomalous behavior.

Of interest are relative works to rupture calculated by Holden for various strain rates, given in Table 6.8.   These values are relative to an interpolated value for a 1 %/sec strain rate.

They can be regarded as only approximate, since they are based on the products of breaking tenacity and extension. Such a calculation, in effect, assumes a strictly linear tenacity-extension curve up to the breaking point. The departures from linearity in the actual curves, however, may not be sufficiently great to introduce much error.

More recently, Krizik, Mellen, and Backer,[42] using a falling-weight and a pneumatic-hydraulic tester, have obtained on yarns of a variety of fiber types, tenacity-extension results similar to those of Meredith, Schiefer and Smith, and Lewis. The Krizik team also studied the effect of different strain rates, up to 3.6×10^5 %/min, on the rupture properties of nylon yarns from webbing exposed to sunlight for various periods of time. Plot-

Table 6.8. Relative Work to Rupture of Various Yarns at Different Strain Rates[37]

Strain Rate (%/sec)	0.033	3.3	4400	66,000
Nylon, "Type 100"	0.80	1.08	0.83	1.05
Nylon, "Type 300"	0.91*	1.12	1.31	1.50
Nylon, undrawn	0.80	1.09	2.71	3.30
Nylon, monofil	0.82	1.07	0.89	1.15
Terylene	1.0	1.0	1.06	1.0
Cellulose acetate	0.98	1.03	1.06	1.03

*Obtained at 0.33 %/sec

ting their results in a manner similar to that of Figure 6.10, they found that, in general, the breaking loads of the exposed yarns increased, and the breaking extensions decreased with increased strain rate. Similar graphs were obtained for nylon thread, tested as a single straight specimen, and tested with two pieces of thread looped through each other, in the manner of two chain links. These plots showed that for both types of specimens, the breaking load rose slightly with strain rate, while breaking extension appeared to be unaffected. They made the further observation that "it appears. . . that the variability of the breaking strength of straight and looped samples is not affected as much by strain rate as is the variability of extension at rupture."

Using conventional quasi-static testers, pneumatic and hydraulic machines, and a ballistic method, such as are described in Chapter 3, Laible and Morgan[43] have studied the dependence of the tenacity-extension relationships of polyvinyl alcohol (vinal) yarn on strain rate. Their experiments were conducted on a

1220-den continuous-filament, oriented sample. The strain rates ranged from 20 to 2×10^6 %/min (0.2 to 7.2×10^5 in./min). The more significant of their findings are graphically summarized in Figure 6.24.   It can be seen that the family of curves generally follows the same pattern, as the strain rate is increased, as has been found for other fiber types.   Breaking energy density was found to remain practically constant (43 joules/g) for all strain rates.

The quasi-static and impact tests were performed with such a surplus of energy in the rupturing machine or missile that there was no sensible loss of velocity during the deformation and rupture process.   Thus, this process could be thought of as being at a constant strain rate. By a procedure similar to that of Morgan and the Celanese group (described in connection with Figure 6.10),

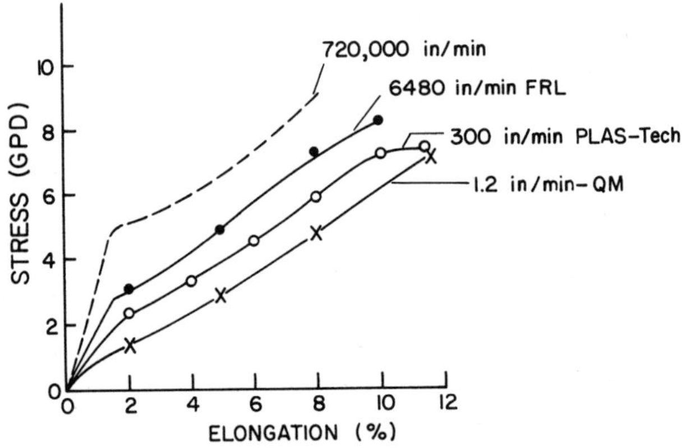

Fig. 6.24. Tenacity(stress)-extension curves of oriented polyvinyl-alcohol yarn at various rates of extension. [After Reference 43].

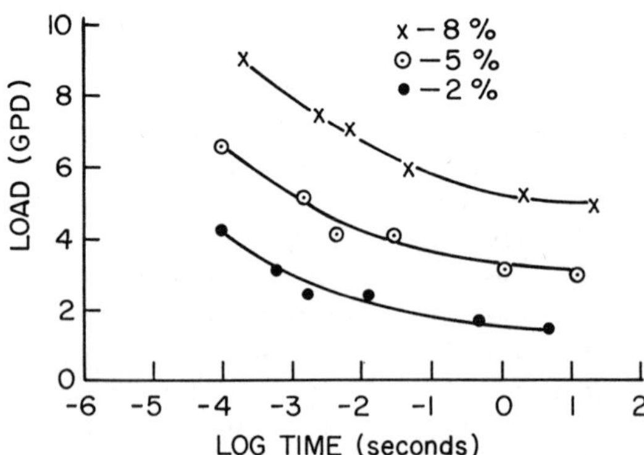

Fig. 6.25. Tenacity(stress) in polyvinyl-alcohol fibers, at three extensions (from Figure 6.24) as a function of time to reach these extensions. [After Reference 43].

the tenacity at a given extension could therefore be plotted as a function of the time required to reach that tenacity (or extension). Such a graph is shown in Figure 6.25 for each of three extensions. Interpreting these results in terms of a distribution of relaxation times, Laible and Morgan concluded that their highly crystalline, oriented vinal sample showed a sizable viscous component in its tensile behavior, and, at even the highest attainable strain velocities, no pure elasticity. This observation appears to be true of textile polymers generally.

6.2.3 Ropes. A comparison of the impact strengths of sisal and nylon climbing ropes was made by Newman and Wheeler, using the falling-weight apparatus previously described.[73]   The specimens were spliced so as to have, at each end, loops that engaged a supporting link and a link affixed to the pan on which the weight fell. The measurement of elongation of the specimen included stretch in the loops, or eyes, at the ends of the specimen. Specimens 5 and 10 ft long were used, and the height of drop was fixed at twice the specimen length. Impact energy was varied by changing the number of discs which, bolted together, formed the falling mass. Impact energy was computed as the product of the weight of this falling mass by the height of drop, including the elongation of the specimen, plus this elongation times the weight of the pan.

Newman and Wheeler were primarily interested in determining the impact energy required to break one strand of the specimen, such a break being taken as the criterion of failure. Hence, the loads used for the falling weight ranged from relatively low ones that merely stretched the specimen in impact, without breaking any strands, through loads that broke one or more, if not all the strands. Each specimen was impacted only once. Quasi-static tests, in which the specimens were stretched to failure at the rate of 1 in./min, were also conducted. From these data the energy required to break one strand of a specimen having the average length of those used in the impact tests was computed.   The results of the tests on the 10-ft specimens are summarized in Table 6.9. The impact velocity was evidently in the neighborhood of 35 ft/sec.

It is evident that these data support the authors' conclusion that "the energy required to cause failure under impact loading was greater than the energy required to cause failure under [quasi-] static loading. The stretch of the ropes at failure was practically the same under impact and static loading." The authors gave the further opinion that the results indicated that energy values computed from quasi-static tests of these ropes gave safe estimates of their performance under impact loading. The results that have been cited for fibers and yarns in the foregoing sections show this opinion to have limited applicability.

Table 6.9. Rupture Properties of Ropes in Quasi-Static and Impact Tests[73]

Sample	Load (lb)	Extension (%)	Breaking Energy* (ft-lb)	Impact Energy (ft-lb)
Sisal, 4 strand, $\frac{9}{16}$-in. diameter	Quasi-static	18.1*	1860	—
	97 to 162	13.8 to 28.4	—	2080 to 3700
Nylon, Type 300 3 strand, $\frac{7}{16}$-in. diameter	Quasi-static	52.7*	6490	—
	268 to 328	45.8 to 51.3	—	6740 to 8570

*Refers to the rupture of one strand.

Table 6.10. Rupture Properties of Ropes of Various Materials in Quasi-Static and Impact Tests[29]

Property	Nylon 66		Dacron	Manila		Steel-Wire Rope
Diameter (in.)	0.5		0.5	0.75		0.25
Specific length (ft/lb)	14.8		12.1	6.0		9.1
Quasi-static breaking force (lb)	7875		6275	5700		8000
Impact Data	Dry	Wet	Dry*	Dry	Wet	—
Breaking energy Density (ft-lb/lb)	15,660	15,000	7300	1800	1120	600
Relative rating**						
For equal weight	870	835	400	100	62	35
For equal diameter	800	750	450	100	62	200
For equal strength	255	245	180	100	62	16

* Dry and wet properties of Dacron are the same.
** Dry manila = 100.

In work conducted for the du Pont Company,[29] a series of impact tests was made comparing nylon 66 and Dacron ropes with ones composed of the natural manila fiber and steel wire. In the falling-weight method that was used, the specimens were 15 ft long with eye splices on each end, as in the Newman and Wheeler tests. The falling weight, ranging from 400 to 900 lb, was dropped from various heights so as to give impact velocities from about 11 to 28 ft/sec. The mechanical properties of the samples, and the breaking energy densities determined in the impact tests are presented in Table 6.10; the particular impact velocity, or velocities, at which these values were obtained was not reported.

The relative ratings of the samples for equal weight of rope (proportional to breaking energy density itself), for ropes of the same diameter, and for ropes of equal quasi-static breaking force,* were computed and are quoted in Table 6.10. The contribution of extensibility to breaking energy is strikingly brought out by comparing the impact data on the nylon and steel wire samples, both having about the same breaking force. The breaking extension of the nylon rope would appear to be 10 to 15 times that of the steel rope.

6.3 Woven Structures

6.3.1 Wide Fabrics.

Relatively few impact measurements have been made on fabrics. Studies of the effect of rate of loading on the strength of balloon fabrics were made by the British Aeronautical Research Committee as far back as World War I.[12] The tests were actually made on specimens of 2-in. width, and not on yarns taken from the fabrics; but the times to rupture were on the order of minutes, so that the rates were not at the impact level. More recently, some insight into the impact behavior of fabrics has been sought in high-speed tests on yarns taken from the fabrics of interest.[42,56] Reference has been made to the results of some of these studies in the foregoing section. Most tensile tests of actual fabrics have been made on webbings, but a few have been made on specimens cut from fabrics of the more common widths.

Impact breaking tests on plain-woven cotton and Dynel fabrics, using a falling pendulum, have been reported by Dickson and Davieau.[27] The tests were made on strips ravelled to 1 in., with a 31-in. length between clamps. The averages of results on five specimens of each sample are given in Table 6.11. As can be seen, measurements were made in both the warp and filling directions. It may be noted that in the cotton fabrics the breaking energies are roughly proportional to the fabric weights. The Dynel fabrics, in comparable weights, must have had breaking energies many times those of the cotton.

---

*Designating the latter parameter thus implies that the force-extension curves are linear, a fact which is only approximately true.

Table 6.11. Impact Rupture of Fabric Strips[27]

Fiber Type	Fabric Weight (oz/yd^2)	Count of Fabric (yarns/in.)		Cotton Count (hanks/lb)		Breaking Energy (in.-lb)	
		W	F	W	F	W	F
Cotton	4.18	71	58	23.7/1	24.7/1	31.4	52.7
Cotton	9.02	30	26	4.8/1	4.8/1	104	108
Dynel	6.16	37	28	16.5/1	8.0/1	353	216
Dynel	10.5	38.3	28.5	10.3/2	10.0/2	*	*

W = warp; F = filling.
* Did not break.

In a study of the translation of yarn into fabric properties, Krizik, Mellen, and Backer conducted impact tests on a lightweight parachute fabric, as well as on the yarns from which they were woven.[42] The interest was in determining what effect rate of extension has on the efficiency with which yarn properties are translated into corresponding properties in a fabric. Two parameters were employed to express the yarn-to-fabric relationship quantitatively. One was load efficiency, which is obtained by dividing the fabric breaking force by the number of yarns broken, and expressing this as a percentage of the breaking force of yarns taken from the fabric in the same direction. The other parameter was strain ratio, which is the breaking extension of the fabric in one direction divided by the breaking extension of yarns in the same direction.

The specimens used were ravelled strips 1 in. wide, with a gage length of 8 in. Tests were repeated with the test length lying in both the warp and filling directions, and at two quasi-static rates of elongation and one impact rate. Five specimens were used for each test. The results are shown in Table 6.12. The authors noted the tendency, shown in these data, for the load efficiency to rise appreciably at the high rate of loading (2.9×10^3 in./min) in the impact test. The yarns used in this fabric were of low twist. The substantial reinforcement afforded the yarns by the weaving, as reflected by the generally high load efficiencies, appears to be even more pronounced under impact loading. The force-extension behavior of a strip (8 in. by 1 in.) of this fabric, in both the warp and filling directions in a low-speed and an impact test is shown in Figure 6.26.

Krizik and associates also studied the behavior of seams under impact loading, as compared with behavior in low-speed tests.

Table 6.12. Rupture Properties of Parachute Fabric and Constituent Yarns, in Quasi-Static and Impact Tests[42]

Yarn Direction	Test Speed (in /min)	Fabric		Yarn		Load Efficiency (%)	Strain Ratio
		Breaking Force (lbs /in)	Breaking Extension (%)	Breaking Force (lb)	Breaking Extension (%)		
Warp	1	43.9	25.5	0.350	23.7	98.7	1.077
	10	46.3	25.4	0.386	23.6	94.8	1.075
	2.9×10^4	62.8	21.8	0.408	24.1	123.0	---
Filling	1	40.7	34.2	0.344	34.0	101.5	1.005
	10	42.3	34.3	0.368	33.1	98.2	1.034
	2.9×10^4	57.8	28.8	0.395	28.0	125.0	---

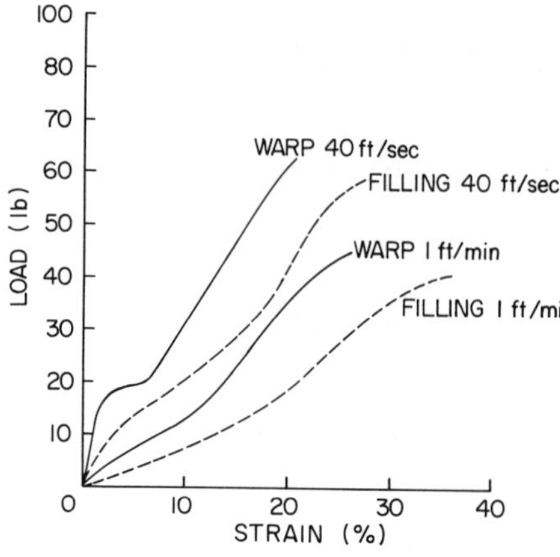

Fig. 6.26. Force (load)-extension curves of parachute fabric, in warp and filling directions, at two rates of elongation. [After Reference 42].

In this work a quantity called <u>seam efficiency</u> was used; this is the breaking force of a fabric strip across a seam, expressed as a percentage of the breaking force of a strip of the sample without a seam. The experiments were conducted on a nylon parachute cloth (1.1 oz, 126×115 construction)[*] with and without seams. The specimen was 2 in. wide, not ravelled, with slits $\frac{1}{2}$ in. long cut on both sides of the seam and parallel to it, on both edges of the strip. The strips not containing seams were similarly notched. This special form of specimen was adopted to avoid abnormal behavior of the seam, leading to very low values. The gage length was again 8 in., with the seam at the middle of the strip. The same three testing speeds mentioned in the previous paragraph were employed in this work. The results, summarized in Table 6.13, show a sharp decrease in seam efficiency as the rate of loading is increased from the quasi-static to impact. The sample without a seam shows the usual (though, in this case, small) increase in breaking force at the higher rates of elongation.

Table 6.13. Effect of Rate of Elongation on Efficiency of Seams in Parachute Fabric[42]

Test Speed (in./min)	Breaking Force (lb)		Seam Efficiency (%)	Strain Ratio
	With Seam	Without Seam		
1	32.0	28.1	114	1.31
10	35.8	34.6	103	1.37
2.9×10^4	26.0	36.8	70	0.81

*Probably the same fabric as that discussed in the preceding paragraph.

6.3.2 Webbing. Stang, Greenspan, and Newman obtained a force-elongation data on two types of nylon and one type of cotton webbing, using falling-weight apparatus equipped with a strain gage.[103] The webbings, all $1\frac{3}{4}$ in. wide, of the type used in parachute harness, consisted of a rib-weave nylon, a herringbone-weave nylon, and a herringbone-weave cotton. The falling-weight used in the impact tests had a constant mass of about 200 lb, while the gage length was about 36 in., in all cases. At least six specimens of each webbing type were tested; in most cases the specimens were not ruptured. Only one specimen of each type of webbing was tested at low speed (4 in./min) for representative quasi-static behavior. The results of these tests are summarized in Figure 6.27. Initial impact velocities were around 24 ft/sec for the nylon samples and around 11 ft/sec for the cotton.

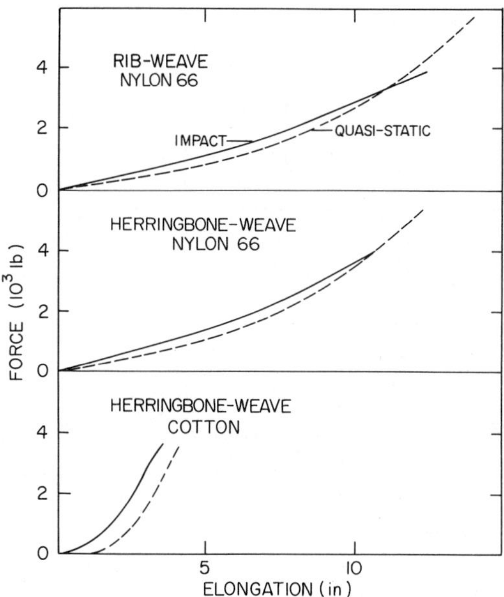

Fig. 6.27. Force-elongation curves of three types of webbing, obtained in quasi-static and impact tests (based on data of Stang and associates[103]).

It can be seen that the force at any elongation is, in general, higher in the impact test than in the quasi-static. This is in line with results that have been cited for single filaments, yarns, and other fabrics. However, except in the case of the cotton webbing, the differences are not great; but, then, the impact velocities were not very high. Obviously, computed energy absorption, at any given elongation, is greater in the impact test than in the low-speed, in all samples. Only three nylon and two cotton specimens were impact loaded to rupture. While the data

Table 6.14. Rupture Properties of Nylon Webbing at Various Impact Velocities[118]

Webbing Sample	Position Number											
	1			2			3			4		
	V	F	E	V	F	E	V	F	E	V	F	E
3600-lb	204	2310	37.3	289	1970	35.4	482	2010	---	726	2410	41.1
6000-lb	216	4560	37.4	280	4560	47.7	507	4570	41.3	750	4110	43.9
9000-lb	221	5890	22.0	298	6340	27.2	516	4810	20.5	737	4540	28.2
10,000-lb	36	6190	24.6	76	7400	30.6	152	5700	31.5	210	6340	31.5

V = average impact velocity (ft/sec).

F = average breaking force (lb).

E = average breaking extension (%).

for the conclusion are thus rather meager, these workers state that the energy absorbed by the broken nylon specimens is more than three times that of the broken cotton specimens.

As an example of the performance of their ballistic equipment, Chu, Coskren, and Morgan[18] have presented curves obtained on a similar structure at 0.5 in./min and 175 ft/sec. These curves also show that the impact force at any selected elongation is higher than the quasi-static. The sample was not identified as to size, weight, or fiber content.

Typical results obtained with the rocket sled described in Chapter 3 have been reported on four nylon webbings, designated as 3600-, 6000-, 9000-, and 10,000-lb (the quasi-static breaking load).[118] For each run V-shaped specimens were mounted at each of the four positions at which the sled had successively higher velocities. The position and velocity of the sled at successive times during a run was determined by means of track-side pickups spaced at 50-ft intervals. Since it was found that in most cases the webbing specimens did not significantly decelerate the sled, it was thus possible to compute the impact velocities at each position. Breaking forces were measured with tensiometers on the sled, as well as with such units attached to the clamps holding the ends of the V-shaped specimens.

In Table 6.14 are summarized the average velocities and breaking forces at each position for a number of runs with each of the four samples. It can be seen that there is no discernible, consistent trend in breaking force with impact velocity. While Williams and Benjamin[118] acknowledge that the data are insufficient to justify a firm conclusion, they tentatively advance the inference that over the velocity range 200 to 750 ft/sec the breaking force is not significantly affected by impact velocity. Table 6.14 would seem to indicate that this range could be extended downward to 30 ft/sec. In any case, these authors speculate that if this inference were substantiated as a fact by additional data, the average breaking force, with an adequate safety factor, could be used as a pertinent parameter in the design of equipment in which webbing is employed. Going further, they call attention to the fact that, for the various types of webbing, the over-all average breaking force was a fairly constant fraction of the nominal quasi-static breaking force: 0.60, 0.74, 0.60, and 0.64 for the successively heavier webbings, respectively.

Also included in Table 6.14 are average breaking extensions based on data from some of the runs in which the breaking forces were obtained. The elongations of the specimens, from which these breaking extensions were computed, were measured as follows: The time from the initial application of force to a specimen to the instant the maximum (breaking) force was reached (a fraction of a second) was determined from oscillograph records

of the tensiometer signals. Multiplying this value by the sled
velocity (taken to be constant) gave the stretch.

The extensions shown in Table 6.14 are in the range of the 30
to 35% reported for the quasi-static breaking extensions, and are
higher than were expected by these workers, on the basis of impact tests on other materials. However, these breaking extensions are not out of line with the force-extension curves (for 36-
in. specimens) shown in Figure 6.27 for nylon webbing. As in the
case of breaking force, no consistent effect of impact velocity
on breaking extension is seen in the extension data in Table 6.14.

A study of some aspects of the mechanics of the static-line
system used to accelerate the deployment and inflation of a parachute canopy was made by Coy.[22] The static lines, 15 ft long,
were formed from heavy nylon webbing by having the edges turned
inward, so as to overlap, and stitching them down the middle,
the full length of the line. As a part of his study, Coy obtained
the impact tensile properties of the static lines at elongation
rates of 25, 30, and 35 ft/sec, using a falling-weight method,
with electrical gages for measuring force and extension.   Two
falling weights of 200 and 300 lb were used, to obtain different
energies at the same impact velocity. Conventional quasi-static
force-extension curves were also obtained.   Only one type of
nylon webbing (Type XIII, Spec. MIL-W-4088) was studied in
these tensile tests.   The results are presented in Figure 6.28.
These curves can be seen to resemble closely those for the nylon webbings in Figure 6.27. Evidently, forming a length of nylon webbing into a static line does not noticeably alter the character of its force-extension curve.

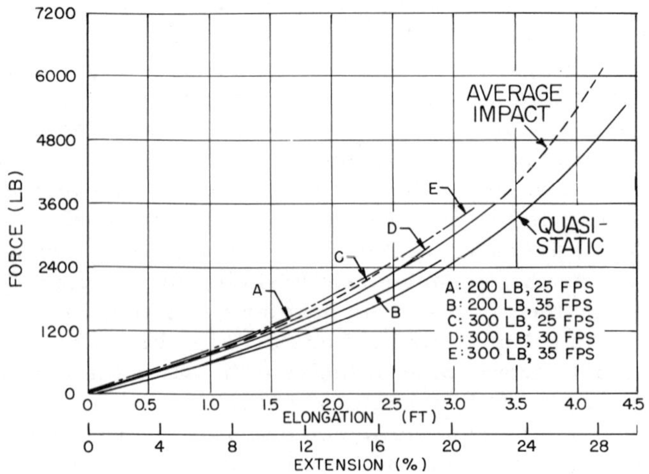

Fig. 6.28. Force-extension curves of a nylon static line, obtained in quasi-static and impact tests. [After Reference 22.]

6.3.3 Studies of Body-Armor Fabric.   Fabrics of numerous
constructions, weights per square yard, and fiber materials
have been evaluated in the U.S. Department of Defense and con-
tractors' laboratories, for use as the protective element in armor clothing, such as the combat soldier's vest. These fabrics
have been tested ballistically in layers ranging from 1 to perhaps 25 or 30, by methods such as those described in Chapter 3.
Typical of the results obtained in a V_{50} test, when the number of
layers is varied, are those on a 2×2, basket-weave nylon 66
armor fabric, of approximately 14 oz/yd^2, similar to the present standard material.[112] Panels were prepared having 2, 4, 8,
10, 12, and 16 fabric layers. These were then tested for the
V_{50} ballistic limit, using the standard procedure, with the results shown graphically in Figure 6.29.*

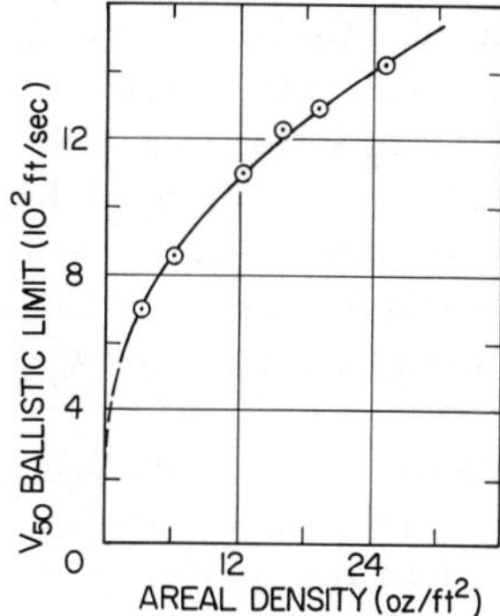

Fig. 6.29.   Ballistic limit of nylon
fabric as a function of the areal density of the panels formed from various numbers of layers of fabric [pri-
vate communication].

It was on an experimental curve such as that shown in Figure
6.29 that the generalized curve (Figure 4.13) proposed for theoretical discussion in Chapter 4 was based. In the region of the
experimental points the curve was found to be well fitted by the
equation

$$V_{50} = 456(\rho h)^{0.357} \tag{6.4}$$

where ρh is the areal density of the test panel, in oz/ft^2. This
empirical equation appears not to be valid below 2 oz/yd^2, giving
V_{50} values lower than those found experimentally.   It will be
recognized as being a different version of the proportionality expressed by Equation 4.34, which could be written:

$$V_{50} = \text{const} \times (\rho h)^{0.5}$$

An extensive microscopic examination of a 12-layer panel
of nylon armor fabric, after it had been subjected to a V_{50} test,

---

*Private communication, Roy C. Laible, Quartermaster R.
and E. Command, Natick, Massachusetts (M.E.R. No. 2365).

was conducted by Susich, Dogliotti, and Wrigley.[106]   While not yielding directly the numerical parameters of a mechanical nature, this study has provided some important insights into the phenomena occurring in fabric structures during the impact process. Photographs, at low magnifications, were obtained of the holes produced in the panel by the impinging missiles. These were views over the faces of the fabric layers, along the surfaces of the fabrics, and of cross sections of the holes, made in both the warp and the filling directions. To obtain the latter, an imbedding technique (involving polymerization), which "froze" the distorted and broken yarns in place, was used, so that the structure would not be disturbed by the cutting edge. At higher magnifications, the character of the mechanical and thermal damage to the individual filaments was studied.

Susich and associates summarize their findings as follows: "The microscopical study revealed permanent deformation and damage in the impacted area. Dislocations, flattening, straightening out, extension, unravelling, and rupture in both yarn systems, and various types of defects of nylon filaments were observed. These consisted of softening, melting accompanied by deorientation, heat relaxation, even decomposition, burning, and fibrillation. . . . These phenomena represent damage because they alter fiber properties essential for protection against the impact of the missile. . . .

"It became evident that the permanent alterations of the fabric due to impact were limited to a region close to the path of the projectile. The impact remains confined in the panel to a relatively small area because the presence of two yarn systems prevents the propagation of the stress wave along the whole yarn length. Apparently cross yarns behave like clamps, limiting the stress to the neighborhood of the impacted spot.   The cross yarn closest to the impact may not act completely; however, the next or the one after does. This restricts the impact to that area of the fabric which was in close contact with the missile.   An exception to this behavior is the slight dissipation of stress in layers exposed to the essentially decelerated missile without appreciable backing. These layers were either insignificantly ruptured or remained entirely unbroken. They did not contribute much to the ballistic resistance of the panel."[106]

Chapter 7

SUMMARY DISCUSSION

7.1 Applications

While the impact and dynamic behavior of fibers, yarns, and fabrics has been fairly well explored and subjected to measurement, there is little information (quantitative, in particular) in the literature about the dynamic performance of the end-use structures composed of these textile elements. Some consideration has been given to the requirements of textile items in particular impact applications, but in almost no case have these been the subject of experimental measurement.

For instance, in discussing the application of a textile material to use in a safety rope, Schiefer and associates[86] brought out the idea that high impact rupture energy should not be the criterion, exclusively. They pointed out that yarns that have high energy recovery, that is, have the ability to store a large portion of the total work done in stopping the fall of a man (where rupture does not occur), would be unsuitable. The reason is that when the fall is finally stopped, such a rope would jerk the man upward, increasing the hazard to him. Schiefer and colleagues go on to say: "A low value for energy recovered . . . indicates that there will be little or no rebound. It also indicates that the material will be permanently stretched and therefore no longer effective for absorbing an impact load. Safety equipment made from such material would be unsuitable for further use after an impact load that utilizes any considerable part of its energy-absorbing ability."

These writers state that impact data on yarns are not adequate for predicting the impact properties of intermediate components, such as ropes and webbings, made from the yarns. Having in mind safety equipment made from rope and webbing, they recommend further that the equipment itself be impact tested. They say: "This is important because in such equipment, points of suspension may deviate critically in their behavior from the rest of the rope." Obviously, in the absence of any soundly based principles on which the impact properties of fibers or yarns can be interpreted for increasingly more complex structures, the precautions have quite broad application.

Further consideration was given to the performance of ropes in safety lines by McCrackin, Schiefer, and colleagues, in

161

another publication.[60] On the basis of the limiting breaking velocity of nylon yarn (200 m/sec), as derived from longitudinal-
impact data, they computed the minimum weight of safety rope
required barely to stop and support a construction worker falling through a specified distance. For a 200-lb worker falling
20 m, this weight of rope was found to be 2 lb. Further consideration revealed that the size of nylon rope which would meet
this minimum requirement would have a quasi-static breaking
force of about 4000 lb. Thus, it was concluded that a rope having a strength 20 times the weight of the worker would provide
no safety factor whatever.   The computation, however, was
based on the assumption that the rope has the same impact behavior as yarn; this assumption, these writers recognized, should
be examined experimentally. Nevertheless, the calculated results indicate the hazards that lie in the uncritical application of
quasi-static parameters to impact situations.

As has been noted, textiles may be, and usually are, subjected
to impact loading in processing, as well as in some of the serv-
ices into which they are put. In a study of the interaction of textile structures and high-speed processing, Morgan[69] analyzed
some results obtained on the weaving operation. An appreciation
of the cyclic impact loading that the yarns undergo can be gained
from his statement: "At a loom speed of 190 picks/min. the frequency of the complete [warp] tension cycle is 3.1 cycles/sec.
However, the time duration of the beat-up force pulse is much
shorter than the complete tension cycle, and has been found to
be approximately 30 millisec. . . . Thus, the beat-up cycle can be
thought of as having a frequency of approximately 33 cycles/sec."

Consideration of the forces involved revealed that the beat-up
action had a tenacity amplitude of about 0.15 g/den, applied in
30 millisec. Morgan further noted: "While the warp tension varied between 20 and 27 gm. [on a 140 den yarn]. . . the corresponding fabric tension on the other side of the beat-up reed varied from 5 to 25 gm., approximately. Thus the tension variations
on the fabric side are more severe than on the warp side."

Morgan also studied the loading cycle on a thread in a high-
speed sewing operation. The motions of the thread and needle as
functions of time through one such cycle are shown in Figure 7.1.
In the operation studied, the sewing rate was 5000 stitches/min
with a maximum needle velocity of 25 ft/sec. At the start of the
cycle shown in Figure 7.1, a particular point on the thread has
just been carried through the fabric by the needle. The thread
is hooked by the bobbin, and its speed is accelerated to an estimated 100 ft/sec downward. The needle, at that instant, is moving upward at 25 ft/sec, so that there is a relative velocity of
125 ft/sec between the needle and the thread.   When the point
(on the thread) that is being followed reaches its lowermost po-

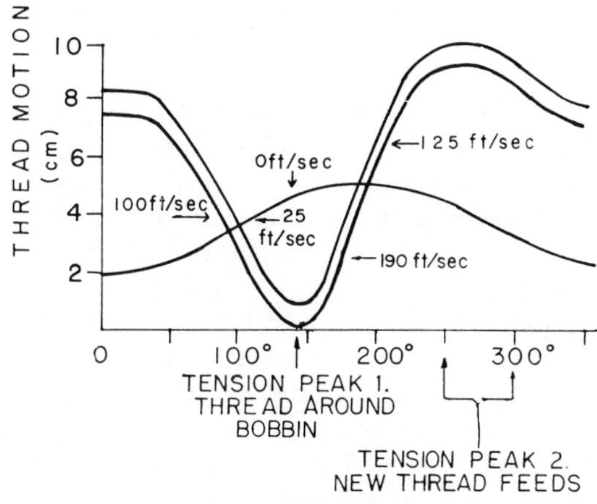

Fig. 7.1. Motion of thread and needle in sewing machine, as a function of time through one cycle.[69]

sition, the take-up mechanism pulls the thread upward at a velocity which very quickly reaches 190 ft/sec. Thus, there is a change in the velocity of the thread of 290 ft/sec in an interval of about 12 millisec. Morgan remarked that the actual upward thread velocity might be even higher than the estimated 190 ft/sec.

That further studies are needed in just this one area of processing and service where textile structures are subjected to high-speed loading, is reflected in Morgan's suggestion that the complex and expensive equipment required be made available to an academic institution, so that a complete investigation of sewing could be undertaken, to the benefit of technology generally.

7.2 Suggested Lines for Future Research

Such parameters as the "snatch" force exerted on a parachute load by the opening canopy, through the suspension lines, have been calculated or measured[16,34] for various conditions. However, in most cases (by far) only the quasi-static mechanical properties of the textile components seem to be taken into consideration in relating these components to the adequacy of the end-item under impact conditions. Indeed, Brown says, "Generally, the elasticity of the fabric or of the cordage in the rigging [suspension] lines has no measurable effect on the ability of the parachute to withstand shock loading." [Reference 16 (p. 127)]. The elasticity to which he refers is undoubtedly that measured in quasi-static tests. Undoubtedly, also, data of this type are used as design criteria because they are the only type that is generally available.

There would appear to be need for research correlating the impact properties: breaking force, extension, and energy, of various textile materials and assemblies, with their performance under impact loading in the end-use structure of which they form a part. It is conceivable that, ultimately, it might be found that materials having superior impact properties not reflected in quasi-static tests could be used with savings in the amounts, and hence, weights, of material required, with no diminutions in safety factors. Alternatively, improvements in safety factors might be the desired achievements. Related new applications involving impact loading, in which textiles could be advantageously used, might suggest themselves in such a study.

The V_{50} test for the ballistic limit of armor fabrics and vests[111] yields data in which the variability, by some standards, might be considered high. A critical review of this test, with respect to experimental design as well as to physical equipment, would probably be profitable, in view of the crucial role played by the V_{50} test in the procurement and acceptance by the U.S. Government of personnel-armor items. Consideration should be given to the possible application of newer statistical techniques.

From the dearth of information on the impact behavior of fabrics, as has been brought out in Chapter 6, there is obviously a need for a broad research program in this area. It is suggested that investigations be made into the influence of yarn geometry (size, twist, and so forth) on the impact properties not only of the yarns themselves, but of the fabrics and other assemblies (such as ropes) that they form. Fabric construction (within the woven and knit categories), as a variable influencing impact behavior, would seem to show much promise for profitable study. A beginning in this direction has been made at the National Bureau of Standards where, under Smith,[93] the influence of yarn structure on the impact properties of the yarn has been studied on a limited number of samples employing a cellulosic fiber.

Further work along the lines pursued by Morgan,[28] Laible and Morgan,[43] and Hall,[35] aimed at relating the impact and dynamic behavior of textile polymers to viscoelastic parameters and molecular structure, should be encouraged. Such research would lead to fuller knowledge of the influence of such characteristics as molecular weight, degree of crystallinity, and orientation, on macroscopic impact properties. The ultimate objective would be to make possible the definitive selection of the most suitable material and the optimum processes of fiber manufacture to achieve maximal performance in particular impact applications.

Further development of the theory of impact phenomena along

the lines initiated at the Chemical Warfare Laboratories,[40,77,78,80] could be profitably pursued. The theory would be brought nearer satisfactory completion with: (a) experimental verification of the strain at particular positions along the yarn, as a function of time, as given by theory; (b) a study of stress-strain behavior at strain rates in the 1 to 5×10^6 %/sec range, to determine whether, as Petterson and colleagues predict, there is a peak breaking stress in this range; and (c) further study of the strain distribution in a yarn after reflection of the strain front.*

The extension of the dynamic theory of impact loading, as developed at the N. B. S. and C. W. L. for linear structures, to transverse impact on two-dimensional structures (fabrics) holds promise of advancing the understanding of the processes involved in the defeat of a missile by body armor.   It is expected that such a theoretical analysis would itself suggest a parallel experimental program.

----

* Smith and associates have recently discussed aspects of some of these subjects.[92,94]

REFERENCES

The following collection contains not only publications to which reference is made in the text of this monograph, but, to make it more nearly a complete bibliography, other literature having a bearing on the subject. Those references that have not been cited in the text are designated by an asterisk (*).   It is believed that the collection will be more valuable for being selective, and hence, only a few items have been taken from the vast literature on the impact testing of metals and other nontextile materials. Similarly, the number of entries pertaining to the various mechanical properties of textiles other than impact has been severely limited. The same is true of those bearing on the extensive technology of textile assemblies that depend, in part, on their impact properties for satisfactory performance. Readers wishing to delve more deeply into these peripheral subjects will find additional references in the listed publications.

1. Alfrey, T., _Mechanical Behavior of High Polymers_, Interscience Publishers, Inc., New York and London (1948).

2. American Society for Testing Materials, "Definitions of Terms Relating to Textile Materials," _ASTM Standards, Part 10,_ 16 (1959).

3. Anon., "FRL Ballistic Range Used to Study Textile Behavior at High Speeds," _Am. Dyestuff Reptr.,_ 50, 69 (1961).

*4. Anon., "Physical Testing Hits High Speed," _Chem. Week,_ 83, No. 26, 76 (1958).

5. Anon., "Bullet-Fast Tests Speed Fiber Research," _Chem. Week, 86,_ No. 3, 55 (1960).

*6. Anon., "New High Speed Single Fiber Tester," _Southern Text. News,_ No. 42, 6 (28 Oct. 1961).

7. Backer, S., and Krizik, J. G., "Problems in Strain Measurement in Impact Tests on Textile Materials," _J. Appl. Polymer Sci., 4,_ 277 (1960).

8. Baker, A., and Swallow, J. E., "Impact Testing of Textile Yarns: Part 1. Experiments with a Falling Weight Method," Technical Note No. CHEM. 1355, [British] Royal Aircraft Establishment (Nov. 1959).

9. Ballou, J. W., and Roetling, J. A., "High Speed Tensile Testing of Fibers," _Textile Res. J.,_ 28, 631 (1958).

10. Balls, W. L., _The Development and Properties of Raw Cotton,_ A. and C. Black, Ltd., London (1915), p. 191.

11. Balls, W. L., Studies of Quality in Cotton, Macmillan and Co., Ltd., London (1928), pp. 228, 258, and 355.

12. Barr, G., "Effect of Rate of Loading on the Apparent Strength of Cotton Balloon Fabric," [British] Aeronautical Research Comm., Reports and Memos., No. 757 (1920).

13. Bellinson, H. R., "Viscose Rayon: Stress-Strain Properties, II. Effect of Rate of Load," Textile Res., 10, 316 (1939).

14. Brinkworth, B. J., "The Propagation of Strain in Textile Cables under Longitudinal Impact," Proc. Conf. on Properties of Materials at High Rates of Strain, Institution of Mech. Engrs., Westminster (1957), p. 184.

15. British Standards Institute, Methods for Testing Textiles, Handbook No. 11, London (1949), p. 138.

16. Brown, W. D., Parachutes, Sir Isaac Pitman & Sons, Ltd., London (1951).

*17. Chaikin, M., and Chamberlain, N. H., "The Propagation of Longitudinal Stress Pulses in Textile Fibers — Parts 1 and 2," J. Textile Inst., 40, T25 (1955).

18. Chu, C. C., Coskren, R. J., and Morgan, H. M., "Investigation of the High Speed Impact Behavior of Fibrous Materials, Part I: Design and Apparatus," WADD Technical Report 60-511, Wright Air Development Div., U.S. Air Force (Sept. 1960).

19. Coskren, R. J., and Chu, C. C., "Investigation of the High Speed Impact Behavior of Fibrous Materials, Part II: Impact Characteristics of Parachute Materials," ASD Technical Report 60-511, PT II, Air Force Systems Command, U.S. Air Force (Feb. 1962).

20. Coskren, R. J., Morgan, H. M., and Chu, C. C., "Behavior of High Strength Parachute Components at Impact Velocities up to 700 Feet per Second," J. Appl. Polymer Sci., 6, 338 (1962).

21. Coskren, R. J., Chu, C. C., and Morgan, H. M., "Impact Performance of High-Strength HT-1 and Nylon Parachute Materials at Mach 0.7," presented at Symposium on Dynamic Behavior of Materials, University of New Mexico, Albuquerque, 27-28 Sept. 1962 (Proceedings to be published by Amer. Soc. Testing Materials).

22. Coy, R. G., "Mechanics of a Parachute Static Line System," WADC Technical Report 55-460, Wright Air Development Center, U.S. Air Force (Nov. 1955).

23. Cross, M. M., and Garrett, D. A., "A High-Speed Fiber Extensometer," J. Textile Inst., 47, T222 (1956).

*24. Dean, Bashford, Helmets and Body Armor in Modern Warfare, Yale University Press, New Yaven (1920).

25. Denham, W. S., and Brash, W., "Instrument for the Ballistic Measurement of Tenacity," J. Textile Inst., 15, T291 (1924).

26. Denham, W. S., and Lonsdale, T., "The Tensile Properties of Silk Filaments," Trans. Faraday Soc., 29, 305 (1933).

27. Dickson, J. B., and Davieau, L. A., "Impact Tester for Textiles,"
    ASTM Bulletin, No. 198, 85 (1954).

28. Dietz, A. G. H., and Eirich, F. R. (eds.), High Speed Testing,
    Vol. I, Interscience Publishers, New York and London (1960),
    pp. 67-96.

29. Du Pont de Nemours, E. I., & Co., "Impact Resistance or Energy
    Absorbing Properties of Ropes of Nylon and 'Dacron'," Technical
    Information Bulletin X-99 (Feb. 1959).

30. Ferry, J. D., Viscoelastic Properties of Polymers, John
    Wiley & Sons, Inc., New York and London (1961).

31. Fiber Society, Symposium on the Impact Properties of Textile Ma-
    terials, New York, N.Y., 6 Sept. 1956:
        Cheatham, J. C., "Impact Properties of Fibrous Materials."
        Coskren, R. J., and Morgan, H. M., "Dynamic Testing of
          Textile Yarns at Ballistic Speeds."
        Ferguson, W. J., and George, W., "Textile Armor Materials."
        Parker, J. P., and Kemic, C. S., "Energy Absorption in Tire
          Yarn During Impact Breaks."
        Smith, J. C., McCrackin, F. L., Stone, W. K., and Schiefer,
          H. F., "Stress-Strain Relationships in Yarns Subjected to
          Rapid Transverse Impact Loading."

32. Foster, G. A. R., "The Effect of the Opening Processes Upon the
    Staple of Cotton," J. Textile Inst., 15, T363 (1924).

33. Garner, Walter, Textile Laboratory Manual, National Trade Press,
    Ltd., London (1949), p. 123.

34. Gimalouski, E. A., "Investigation of Impact Load Absorption Through
    Suspension Line Elongation," Contract No. AF 33 (038) 10401, Air
    Material Command, U.S. Air Force (June 1951); also available as
    WADC Technical Report 52-57, Wright Air Development Center,
    U.S. Air Force (1952).

35. Hall, I. H., "The Effect of Temperature and Strain Rate on the
    Stress-Strain Curve of Oriented Isotactic Polypropylene," J.
    Polymer Sci., 54, 505 (1961).

36. Hindman, H., and Krook, C. M., "The Electric Strain Gage and Its
    Use in Textile Measurements," Textile Res. J., 15, 233 (1945).

37. Holden, G., "Tensile Properties of Textile Fibers at Very High
    Rates of Extension," J. Textile Inst., 50, T41 (1959).

*38. Hsiao, C. C., "Time-Dependent Tensile Strength of Solids,"
    Nature, 186, 535 (1960).

39. Jameson, J. W., Stewart, G. M., Petterson, D. R., and Odell,
    F. A., "Dynamics of Body Armor Materials under High Speed Im-
    pact; Part V. Moving Image Photography for Determining Strain
    History in Yarn," Chemical Warfare Laboratories Report No.
    2185 (Oct. 1957).

40. Jameson, J. W., Stewart, G. M., Petterson, D. R., and Odell, F. A., "Dynamic Distribution of Strain in Textile Materials under High-Speed Impact. Part III: Strain-Time-Position History in Yarns," Textile Res. J., 32, 858 (1962).

41. Kragelsky, E. V., "Physical Properties of Bast Materials," Moscow (1937) [cited by Leaderman].

42. Krizik, J. G., Mellen, D. M., and Backer, S., "Dynamic Testing of Small Textile Structures and Assemblies," Final Report, Contract No. DA-19-129-QM-1308, Quartermaster R. and D. Center (Jan. 1961).

43. Laible, R. C., and Morgan, H. M., "The Viscoelastic Behavior of Oriented Polyvinyl Alcohol Fibers at Large Strains," J. Polymer Sci., 54, 53 (1961).

*44. Laible, R. C., and Morgan, H. M., "The Viscoelastic Behavior of Isotactic Polypropylene Fibers," J. Appl. Polymer Sci., 6, 269 (1962).

45. Lang, W. R., "Ballistic Determination of Work of Rupture of Short Textile Fibers," J. Textile Inst., 42, T314 (1951).

46. Leaderman, H., "Impact Testing of Textiles," Textile Res., 13, No. 8, 21 (1943).

47. Lester, J. H., "Scientific Instruments for Testing Textiles," J. Textile Inst., 1, 63 (1910).

48. Lewis, G. M., "A Method of Investigation of the Stress-Strain Behavior of Fibres at Very High Rates of Extension," Proc. Conf. on Properties of Materials at High Rates of Strain, Institution of Mech. Engrs., Westminster (1957), p. 194.

49. Lyons, W. J., "Dynamic Properties of Filaments, Yarns and Cords at Sonic Frequencies," Textile Res. J., 19, 123 (1949).

50. Lyons, W. J., and Prettyman, I. B., "Method for the Absolute Measurement of Dynamic Properties of Linear Structures at Sonic Frequencies," J. Appl. Phys., 19, 473 (1948).

51. Lyons, W. J., and Prettyman, I. B., "Use of the Ballistic Pendulum for Impact Testing of Tirecord," Textile Res. J., 23, 917 (1953).

52. Maheux, R. C., Stewart, G. M., Petterson, D. R., and Odell, F. A., "Dynamics of Body Armor Materials Under High Speed Impact; Part I. Transient Deformation, Rate of Deformation, and Energy Absorption in Single and Multilayer Armor Panels," Chemical Warfare Laboratories Report No. 2141 (Oct. 1957).

53. Mann, H. C., "High-Velocity Tension-Impact Tests," Amer. Soc. Testing Mater. Proc., 36, 85 (1936)

54. Mann, J. C., and Peirce, F. T., "The Time Factor in Hair Testing," J. Textile Inst., 17, T82 (1926).

55. Massachusetts Institute of Technology, "Impact Behavior of Mechanical Textile Materials," Reports Nos. 1, 2, and 3, and Final Report, Contract DA-19-129-QM-529.

56. Massachusetts Institute of Technology, "Impact Behavior of Textile Materials," First to Sixth Quarterly Reports, and Final Report, Contract DA-19-129-QM-880.

57. Massachusetts Institute of Technology, "Dynamic Testing of Small Textile Structures and Assemblies," First Quarterly Report, Contract DA-19-129-QM-1308; see Reference 42 (Final Report).

*58. McCrackin, Frank L., "Impact Tests of Textile Filaments," Dissertation Abstracts, 17, No. 10 (1957).

59. McCrackin, Frank L., "Effect of Air Drag on the Motion of a Filament Struck Transversely by a High-Speed Projectile," J. Research Natl. Bur. Std., 66C, 317 (1962).

60. McCrackin, F. L., Schiefer, H. F., Smith, J. C., and Stone, W. K., "Stress-Strain Relationships in Yarn Subjected to Rapid Impact Loading: 2. Breaking Velocities, Strain Energies, and Theory Neglecting Wave Propagation," J. Research Natl. Bur. Std., 54, 277 (1955); Textile Res. J., 25, 529 (1955).

*61. Melliker, J. B., and Gailus, W. J., "The Light-Armor Testing Laboratory and Research Relating Thereto," Final Report on QMC 30-F, vol. II; NRC Committee on Quartermaster Problems, Contract No. W-44-109-qm-305 (Dec. 1945).

62. Meredith, R., "The Effect of Rate of Extension on the Strength and Extension of Cotton Yarns," J. Textile Inst., 41, T199 (1950).

63. Meredith, R., "The Effect of Rate of Extension on the Tensile Behavior of Viscose and Acetate Rayons, Silk, and Nylon," J. Textile Inst., 45, T30 (1954); Shirley Inst. Mem., 26, 247 (1952-53).

*64. Meredith, R. (ed.), Mechanical Properties of Textile Fibers, Interscience Publishers, Inc., New York (1956), pp. 77, 184, and 275.

*65. Meredith, R., and Hearle, J. W. S. (eds.), Physical Methods of Investigating Textiles, Textile Book Publishers, Inc., New York, (1959), p. 224.

66. Midgley, E., and Peirce, F. T., "Tensile Tests for Cotton Yarns, II. The Ballistic Test for Work of Rupture," J. Textile Inst., 17, T317 (1926).

67. Midgley, E., and Peirce, F. T., "Tensile Tests for Cotton Yarns, III. The Rate of Loading," J. Textile Inst., 17, T330 (1926).

*68. Morgan, H. M., "Behavior of Textiles Under Impact Loading," presented at Parachute Technology Program, Massachusetts Institute of Technology, 20 June - 1 July, 1955.

69. Morgan, H. M., "The Interaction of Textile Structures and High-Speed Textile Processes," Trans. ASME, 82, Ser. B, 154 (1960).

70. Morgan, H. M., Coskren, R. J., Sprague, B. S., and Singleton, R. W., "High-Speed Tensile Testing of Single Fibers," presented at Fiber Society Meeting, West Point, N. Y., 12 Oct. 1961; also Singleton, R. W., and Sprague, B. S., presented at A.S.T.M., Committee D-13 Meeting, New York, N. Y., 16-19 Oct. 1962 (revised paper to be submitted to Textile Res. J., 1963).

71. Morton, W. E., and Turner, A. J., "The Results of Various Strength Tests on [Mercerized] Fabrics," J. Textile Inst., 19, T189 (1928).

*72. National Bureau of Standards, "Critical Velocity and Stress-Strain Properties of Organic Materials at Very High Rates of Straining," Progress Reports Nos. 9 A115, 3809, 4106, and 4244 (Serial Nos. 1 to 4), NBS Project 0702-10-3841 (1954-55).

73. Newman, S. B., and Wheeler, H. G., "Impact Strength of Nylon and Sisal Ropes," J. Research Natl. Bur. Std., 35, 417 (1945).

74. Parker, J. P., and Kemic, C. S., "A Flywheel Impact Test for Measuring Physical Properties of Tire Cords," J. Appl. Polymer Sci., 6, 255 (1962).

75. Peirce, F. T., "Tensile Tests for Cotton Yarns, IV. The Dynamics of Some Testing Instruments," J. Textile Inst., 17, T342 (1926).

76. Peirce, F. T., "Some Problems of Textile Testing," J. Textile Inst., 18, T475 (1927).

77. Petterson, D. R., and Stewart, G. M., "Dynamics of Body Armor Materials Under High Speed Impact; Part IV. Nominal Dynamic Stress-Strain Curves," Chemical Warfare Laboratories Report No. 2184 (Oct. 1957).

78. Petterson, D. R., and Stewart, G. M., "Dynamic Distribution of Strain in Textile Materials Under High-Speed Impact; Part II: Stress-Strain Curves from Strain-Position Distributions," Textile Res. J., 30, 422 (1960).

79. Petterson, D. R., Stewart, G. M., Maheux, R. C., and Odell, F. A., "Dynamics of Body Armor Materials Under High Speed Impact; Part III. Dynamic Strain-Position Distributions of Nylon Yarn Impacted Transversely," Chemical Warfare Laboratories Report No. 2161 (Oct. 1957).

80. Petterson, D. R., Stewart, G. M., Odell, F. A., and Maheux, R. C., "Dynamic Distribution of Strain in Textile Materials Under High-Speed Impact; Part I: Experimental Methods and Preliminary Results on Single Yarns," Textile Res. J., 30, 411 (1960).

81. Pickard, R. H., and Wallace, W. McG., "Mechanical and Physical Tests for Textile Fabrics," J. Textile Inst., 10, 240 (1919).

82. Pilsworth, M. N., Jr., "Tensile Impact on Rubber and Nylon," Bull. Amer. Phys. Soc., 7 (Ser. II), 241 (1962); also Technical Report PR-3, Quartermaster Research and Engineering Command, U.S. Army (May 1962).

83. Reeves, E. V., "Impact Testing of Yarns and Cords," Textile Weekly, 60, No. 1, 133-134, 137-138 (1960).

84. Rogers, Marguerite M., "Dissipation of Projectile Energy on Impact," Technical Notes Nos. 1-55-P2 and 6-55-P3, University of South Carolina, Contract DA-19-129-QM-128 (Dec. 1954 and May 1955).

85. Sayre, M. F., and Werring, W. W., "Symposium on Impact Testing," Amer. Soc. Testing Mater. Proc., 38, 21 (1938).

86. Schiefer, H. F., Appel, W. D., Krasny, J. F., and Richey, G. C., "Impact Properties of Yarns Made from Different Fibers," Textile Res. J., 23, 489 (1953).

87. Schiefer, H. F., Smith, J. C., McCrackin, F. L., and Stone, W. K., "Stress-Strain Relationships in Yarns Subjected to Rapid Impact Loading," ASTM Special Technical Publication No. 176, 126 (1955); also ASTM Special Technical Publication No. 185, 47 (1955).

*88. Schiefer, H. F., Smith, J. C., McCrackin, F. L., Stone, W. K., and Towne, K. M., "Stress-Strain Relationships in Yarns Tested at Rates of Straining up to a Million Per Cent per Minute," Proc. Intnl. Wool Textile Res. Conf. Australia, vol. D, Melbourne, C.S.I.R.O. (1956), p. 148.

89. Schwarz, E. R., "Samuel Slater Memorial Textile Laboratory," Textile Res. J., 15, 33 (1945).

90. Smith, H. DeW., "Textile Fibers: An Engineering Approach to Their Properties and Utilization," Amer. Soc. for Testing Materials, Philadelphia (1944); also Amer. Soc. Testing Mater. Proc. 44, 543 (1944).

91. Smith, J. C., Blandford, J. M., and Schiefer, H. F., "Stress-Strain Relationships in Yarns Subjected to Rapid Impact Loading. Part VI: Velocities of Strain Waves Resulting from Impact," Textile Res. J., 30, 752 (1960).

92. Smith, J. C., Blandford, J. M., and Towne, K. M., "Stress-Strain Relationships in Yarns Subjected to Rapid Impact Loading. Part VIII: Shock Waves, Limiting Breaking Velocities, and Critical Velocities," Textile Res. J., 32, 67 (1962).

93. Smith, J. C., Blandford, J. M., Shouse, P. J., and Towne, K. M., Stress-Strain Relationships in Yarns Subjected to Rapid Impact Loading. Part IX: Effect of Yarn Structure," Textile Res. J., 32, 472 (1962).

94. Smith, J. C., Fenstermaker, C. A., and Shouse, P. J., "Behavior of Filamentous Materials Subjected to High-Speed Tensile Impact," presented at Symposium on Dynamic Behavior of Materials, University of New Mexico, Albuquerque, 27-28 Sept. 1962 (Proceedings to be published by Amer. Soc. Testing Materials).

95. Smith J. C., Fenstermaker, C. A., and Shouse, P. J., "Stress-Strain Behavior of Textile Yarns Impacted by Rifle Bullets," presented at Fiber Society Meeting, Boston, Mass., 11 Oct. 1962. (submitted 1963 for publication in Textile Res. J.).

96. Smith, J. R., McCrackin, F. L., and Schiefer, H. F., "Stress-Strain Relationships in Yarns Subjected to Rapid Impact Loading: 3. Effect of Wave Propagation," J. Research Natl. Bur. Std., 55, 19 (1955); Textile Res. J., 25, 701 (1955).

97. Smith, J. C., McCrackin, F. L., and Schiefer, H. F., "The Impact-Absorbing Capacity of Textile Yarns," ASTM Bulletin, No. 220, 52 (1957).

98. Smith, J. C., McCrackin, F. L., and Schiefer, H. F., "Stress-Strain Relationships in Yarns Subjected to Rapid Impact Loading: 5. Wave Propagation in Long Textile Yarns Impacted Transversely," J. Research Natl. Bur. Std., 60, 517 (1958); Textile Res. J., 28, 288 (1958).

*99. Smith, J. C., McCrackin, F. L., and Schiefer, H. F., "Characterization of the High-Speed Impact Behavior of Textile Yarns," J. Textile Inst., 50, T55 (1959).

100. Smith, J. C., McCrackin, F. L., Schiefer, H. F., Stone, W. K., and Towne, K. M., "Stress-Strain Relationships in Yarns Subjected to Rapid Impact Loading: 4. Transverse Impact Tests," J. Research Natl. Bur. Std., 57, 83 (1956); Textile Res. J., 26, 821 (1956).

101. Smith, J. C., Shouse, P. J., Blandford, J. M., and Towne, K. M., "Stress-Strain Relationships in Yarns Subjected to Rapid Impact Loading. Part VII: Stress-Strain Curves and Breaking Energy Data for Textile Yarns," Textile Res. J., 31, 721 (1961); ibid., 33, 168 (1963) (letter).

*102. Spȧth, W., Impact Testing of Materials (rev. English transl.), Gordon and Breach, New York (1961), p. 66 ff.

103. Stang, A. H., Greenspan, M., and Newman, S. B., "Dynamic Tensile Tests of Parachute Webbing," J. Research Natl. Bur. Std., 36, 44 (1946).

104. Stone, W. K., Schiefer, H. F., and Fox, G., "Stress-Strain Relationships in Yarns Subjected to Rapid Impact Loading: 1. Equipment, Testing Procedure, and Typical Results," J. Research Natl. Bur. Std., 54, 269 (1955); Textile Res. J., 25, 520 (1955).

105. Supnik, R. H., and Silberberg, M., "Properties of Foams and Laminates Under Shock Loading," Soc. Plastics Eng. J., 15, No. 1 (1959).

106. Susich, G., Dogliotti, L. M., and Wrigley, A. S., "Microscopical Study of a Multilayer Nylon Body Armor Panel After Impact," Textile Res. J., 28, 361 (1958).

*107. Tatnall, F. G., "Speed of Testing: a Summary," ASTM Bulletin, No. 161, 23 (1949).

108. Tipton, H., "Dependence of the Dynamic Properties of Textile Yarns on Strain and Strain Rate," Proc. Conf. on Properties of Materials at High Rates of Strain, Institution of Mech. Engrs., Westminster, (1957).

109. Turner, A. J., and Venkataraman, V., "A Study of Comparative Results for Lea, Single Thread, and Ballistic Tests on Yarns from Standard Indian Cottons," J. Textile Inst., 22, T197 (1931).

*110. U.S. Air Force, Wright Air Development Center, "USAF Parachute Handbook," WADC Technical Report 55-265 (Dec. 1956).

111. U.S. Army, Ordinance Corps, Aberdeen Proving Ground, "Operating Procedure for the Ballistic Acceptance Testing of Vest, Armor...." (1953).

112. U.S. Army, Quartermaster Research and Engineering Command, "MIL-C-12369B (QMC), Cloth, Nylon, Ballistic, for Armor" (1957).

113. U.S. Army, Quartermaster Research and Engineering Command, "MIL-A-12370C (QMC), Armor, Body, Fragmentation Protective, M-1952-A" (1960).

114. U.S. Dept. of Defense, Symposium on Personnel Armor; Naval Research Lab., 4-5 Oct. 1961:
    Smith, J. C., "Characterization of Textile Yarns for Use Under Ballistic Impact Conditions."
    Morgan, H. M., "Dynamic Behavior of Textile Fibers and Structures as Related to Personnel Armor."
    Fugelso, L. E., "Theoretical Study of Penetration and Residual Velocities."
    Corcoran, J. W., "Method of Obtaining Yield Stresses at High Strain Rates."
    Laible, R. C., "Dynamic Properties of High Tenacity Yarns and Their Relationship to Ballistic Resistance."
    Jaskowski, M. C., "Buoyant Insulating Body Armors from Staple Fibers."
    Barron, E. R., "Development of QMC Composite Armor Vest."
    Maisel, H., Chandler, W., and DeCarlo, G., "A Set of Angles of Obliquity for Use in Assessing Body Armor."

115. von Karman, T., and Duwez, P., "The Propagation of Plastic Deformation in Solids," J. Appl. Phys., 21, 987 (1950).

116. Wegener, W., "Dependence of Stress in Synthetic Textiles on Time and Strain," Melliand Textilber., 30, 90, 138, 184, 282, 388, 443, 501, 558 (1949). In German.

117. Wegener, W., and Geuthe, K., "The Constant-Rate-of-Elongation and Free-Fall Methods for Determining the Elongation-Tension Curves of Silk and Synthetic Fibers," Melliand Textilber., 33, 130, 234 (1952). In German.

118. Williams, R. B., and Benjamin, R. J., "Analysis of Webbing Impact Data and Determination of Optimum Instrumentation to Be Used in Conjunction with the Impacting of Webbing," WADC Technical Report 59-694, Wright Air Development Div., U.S. Air Force (Mar. 1960).

119. Wood, G. C., and Chamberlain, N. H., "The Relaxation of Stretched Animal Fibers; 1. An Apparatus for the Determination of Relaxation Curves," J. Textile Inst., 45, T147 (1954).

INDEX[*]

177